Lecture Notes in Networks and Systems

Volume 1366

The series "Lecture Notes in Networks and Systems" publishes the latest developments in Networks and Systems—quickly, informally and with high quality. Original research reported in proceedings and post-proceedings represents the core of LNNS.

Volumes published in LNNS embrace all aspects and subfields of, as well as new challenges in, Networks and Systems.

The series contains proceedings and edited volumes in systems and networks, spanning the areas of Cyber-Physical Systems, Autonomous Systems, Sensor Networks, Control Systems, Energy Systems, Automotive Systems, Biological Systems, Vehicular Networking and Connected Vehicles, Aerospace Systems, Automation, Manufacturing, Smart Grids, Nonlinear Systems, Power Systems, Robotics, Social Systems, Economic Systems and other. Of particular value to both the contributors and the readership are the short publication timeframe and the world-wide distribution and exposure which enable both a wide and rapid dissemination of research output.

The series covers the theory, applications, and perspectives on the state of the art and future developments relevant to systems and networks, decision making, control, complex processes and related areas, as embedded in the fields of interdisciplinary and applied sciences, engineering, computer science, physics, economics, social, and life sciences, as well as the paradigms and methodologies behind them.

Indexed by SCOPUS, EI Compendex, INSPEC, WTI Frankfurt eG, zbMATH, SCImago.

All books published in the series are submitted for consideration in Web of Science.

For proposals from Asia please contact Aninda Bose (aninda.bose@springer.com).

Sonya M. Sultanova · Nilufar U. Babakhanova ·
Said S. Shaumarov · Ma'sud N. Masharipov
Editors

Sustainable Development of Transport

Economy, Transformation, Logistics and ESG Agenda. Volume 2

Springer

Editors
Sonya M. Sultanova
Tashkent State Transport University
Tashkent, Uzbekistan

Nilufar U. Babakhanova
Tashkent State Transport University
Tashkent, Uzbekistan

Said S. Shaumarov
Vice Rector for Research and Innovation
Tashkent State Transport University
Tashkent, Uzbekistan

Ma'sud N. Masharipov
Tashkent State Transport University
Tashkent, Uzbekistan

ISSN 2367-3370 ISSN 2367-3389 (electronic)
Lecture Notes in Networks and Systems
ISBN 978-3-031-88845-8 ISBN 978-3-031-88846-5 (eBook)
https://doi.org/10.1007/978-3-031-88846-5

This Springer imprint is published by the registered company Springer Nature Switzerland AG
The registered company address is: Gewerbestrasse 11, 6330 Cham, Switzerland

Preface

The second volume of the book incorporates the latest scientific findings on the sustainable development of transport and logistics systems, including the transport organizations within them, in accordance with the ESG agenda. The central idea of this volume, which permeates all its chapters, is that transport sustainability entails the comprehensive implementation and equal importance of Environmental (E), Social (S), and Corporate Governance (G) management for the holistic optimization of the impacts of transport and logistics services.

The second volume explores the socio-ecological aspect of sustainability within transport and logistics systems. It examines these systems from the perspective of governmental bodies that manage them, considering their interests in the regions where they operate alongside the interests of residents. This dimension emphasizes that for transport and logistics services to be sustainable, they must meet the population's needs, remain economically viable, and align with the socio-ecological priorities of the meso-level economic environment in which they are provided.

The interests of governmental entities managing transport and logistics systems in relation to the territories in which they operate include the optimization of regional and urban traffic flows and the greening of transport services provided within these areas. The interests of the residents of the territories where transport and logistics systems are located encompass the support of employment and the openness of transport organizations for communication with all stakeholders.

The first part of this volume is dedicated to managing the sustainable development of territorial transport and logistics systems in accordance with the ESG agenda. The relevance of this part to the topic of the volume is justified by its thorough examination of the environmental impacts of transportation use and the provision of scientific and methodological recommendations for mitigating these negative effects. The provided recommendations entail the implementation of green transport technologies and the transition of transport and logistics systems to environmentally clean energy.

As a result, this part delves deeply into cutting-edge international experiences, uncovers underlying patterns, and develops practical solutions for reorienting transport organizations toward energy sourced from renewables, whose generation and use are associated with lower carbon emissions and reduced other waste that harms the environment. Thus, this part shapes the scientific understanding of the ecological sustainability of transport and logistics systems.

Another important topic addressed in this part is the optimization of urban transport and logistics networks and flows. The chapters in this part present enhanced models for organizing urban transport and logistics systems in such a way that these systems fully meet the transportation needs of city residents while avoiding the overloading of urban transport networks.

This is particularly relevant for transit cities through which major logistics routes pass. Although additional revenue from transport transit flows into the municipal budget, transport infrastructure becomes overloaded, reducing the quality of transport services for city residents and deteriorating the urban environment. To promote the sustainable development of urban transport and logistics systems, this part includes recommendations for routing major logistics corridors around cities while maintaining their overall directions.

The second part of this volume systematizes digital technologies and the infrastructural support necessary for the sustainable development of transport and logistics systems. This part expands on the volume's topic by exploring the potential to enhance the resilience of regional transport and logistics systems by creating smart cities. The scientific and practical recommendations provided leverage artificial intelligence to optimize the planning and organization of urban logistics and strengthen environmental oversight of transport and logistics organizations through automation.

Cities are examined as open systems that interact with their external environment in a globalized context. This approach highlights the sustainability of urban transport and logistics systems by emphasizing their integration into the global economy through the promotion of cross-border connections. The chapters offer recommendations for increasing the tourist appeal of regions by sustainably developing their transport and logistics systems.

Additionally, this part contributes to the scholarly exploration of the second part of the volume by introducing the concept of Logistics 4.0 to support the sustainable development of transport and logistics systems. The advantages of Logistics 4.0 include accurate forecasting of logistics route loads and enhanced capabilities for their modeling and optimization from economic and environmental perspectives. This enables the application of Logistics 4.0 solutions in municipal transport management and the corporate governance of transport organizations.

The third part of this volume centers on social communications within transport logistics systems for their sustainable development in accordance with the ESG agenda. The importance of this part to the volume's topic is underscored by its justification of the critical role that human resources play in the sustainable development of transport and logistics systems. It addresses the potential for leveraging local human resources in the activities of transport organizations and highlights the limitations and challenges associated with attracting foreign specialists.

This part is also significant for this volume of the book because it develops a scientific concept of sustainable communications in transport logistics, encompassing social, labor, and marketing communications. Consequently, it presents scientific advancements aimed at ensuring sustainable employment within the transport sector and enhancing the marketing strategies of transport organizations. The marketing recommendations will enable the adaptation of the ESG agenda to the specific consumer preferences of different regions, thereby amplifying the social benefits derived from sustainable transport development. Additionally, this will increase the social support for the activities of transport organizations within their respective territories.

Thus, all chapters of the second volume examine transport logistics systems inextricably linked to the territories they traverse and the socio-economic and natural environments in which these systems operate. The collective scientific and practical recommendations offered in the chapters of the second volume facilitate the achievement of a sustainable equilibrium between transport logistics systems and their meso-level environments, ensuring the balanced development of the transport economy alongside social progress.

Tashkent, Uzbekistan

Sonya M. Sultanova
sofsul@mail.ru

Nilufar U. Babakhanova
lunalika1@mail.ru

Said S. Shaumarov
shaumarov_s@tstu.uz

Ma'sud N. Masharipov
masudcha@mail.ru

Introduction

The second volume of the book is dedicated to the study and scientific pursuit of solutions to the pressing issue of sustainable development in transport and logistics systems operating within specific territories. In this context, the sustainability of a transport and logistics system signifies its harmonious integration into the socio-economic and natural environments in which it exists. The sustainability of transport organizations within these systems is understood as their conscientious stewardship of the environment, care for employees, and support for enhancing the quality of life in the regions where they operate.

Moreover, this volume considers the contemporary context of sustainable transport logistics development, characterized by two significant transformations. The first, ecological transformation, involves intensifying the global fight against climate change and the vigorous decarbonization of the economy. In light of this ecological transformation of economic systems, sustainable transport development requires reducing its carbon footprint and transitioning to green energy sources.

The second transformation, digital transformation, is evident in the Fourth Industrial Revolution. Consequently, sustainable transport development entails the continuous adoption of emerging advanced technologies and their utilization to enhance the economic efficiency and environmental friendliness of transport organizations and foster growth and engagement in addressing socially significant issues within the territories they serve. The ESG agenda encompasses the Sustainable Development Goals (SDGs) and the strategic guidelines for business management aimed at their practical implementation: Environmental (E), Social (S), and Governance (G).

The scientific novelty of the second volume lies in its proposal of a systemic approach to managing the sustainable development of transport and logistics systems and the transport organizations within them in accordance with the ESG agenda. This developed approach includes theoretical foundations, fundamental principles, and methodological recommendations for aligning transport and logistics systems and transport organizations with the ESG framework.

The empirical significance of the second volume lies in its inclusion of applied research that highlights international experiences in sustainable transport development in accordance with the ESG agenda, as well as the innovations for greening urban logistics facilities. This volume proposes social partnership as a cutting-edge tool for ensuring the sustainable development of transport organizations. A dedicated chapter reinterprets the impact of the COVID-19 pandemic on transport logistics from the perspective of its sustainable development.

The target audience for the second volume includes members of the academic community who study sustainable transport development from the viewpoints of territorial, green, and digital economies, as well as labor and education economics. The book elucidates contemporary trends in sustainable transport development and presents advanced scientific research supporting this development amidst global ecological and digital transformations.

An additional target audience for this second volume comprises public administration officials in cities who will find practical recommendations for the sustainable development of urban transport and logistics systems. Representatives of transport organizations will discover practical guidelines for enhancing the management of their marketing communications with stakeholders in the interest of sustainable development, as well as for implementing the ESG agenda in providing transport and logistics services.

Tashkent, Uzbekistan

Sonya M. Sultanova
sofsul@mail.ru

Nilufar U. Babakhanova
lunalika1@mail.ru

Said S. Shaumarov
shaumarov_s@tstu.uz

Ma'sud N. Masharipov
masudcha@mail.ru

Contents

Managing the Sustainable Development of Regional Transport and Logistics Systems in Line with the ESG Agenda

Greening City Streets and Roads in Modern Urban Conditions

Mokhichekhra Z. Ergashova and Temur Sh. Bobonazarov

Abstract The research addresses the issues related to the formation of Tashkent's greening system and air quality in the city, as reported by the IQ Air platform. It offers conclusions and recommendations for improvement and development. The most important characteristic of the urban environment is the microclimate, the condition of which is mainly determined by anthropogenic factors in the environment and, primarily, its pollution. The concept of the master plan of the city of Tashkent determined that the air temperature in the existing green zones is 5° lower than in sparse areas and streets. The authors compared the air temperature inside and outside the car. Elements of microclimatic efficiency and external improvement of green spaces in the conditions of overheating of the suburbs are systematized by the authors. The shortage of parking spaces in Tashkent has led to various problems, with cars frequently parked along the edges of city streets and roads. To mitigate these issues and help moderate the city's climate, the construction of ecological parking lots is recommended. These environmentally friendly parking solutions are widely used around the world. Their primary purpose is to protect the ecological environment while ensuring smooth traffic flow. In these parking lots, special lawn fences are installed to preserve the grass cover, preventing its damage and degradation.

Keywords Urban planning · Greening of streets and roads · Unusual weather conditions · Photosynthesis · Carbon dioxide · Radiation regime · Urban heat island · Road map · Ecological parking · Lawn fences

JEL codes P25 · Q57

M. Z. Ergashova (✉) · T. Sh. Bobonazarov
Tashkent State Transport University, Tashkent, Uzbekistan
e-mail: mohichexra1995@gmail.com

S. M. Sultanova et al. (eds.), *Sustainable Development of Transport*, Lecture Notes in Networks and Systems 1366, https://doi.org/10.1007/978-3-031-88846-5_1

1 Introduction

The political and spiritual progress in Uzbekistan, along with reforms aimed at democratizing society, and the country's socio-economic and scientific-technical advancements, have created favorable conditions for the rapid growth of creative works in the field of urban planning [1]. The design and maintenance of urban green spaces should consider such parameters as the city's population, environmental conditions, and climate [2].

Several laws were enacted regarding the greening of highways and city streets, involving private businesses in these efforts. Local and international expertise played a significant role in the process. High-quality ornamental trees and shrubs adapted to local conditions were planted. As a result, new approaches to greening highways and streets were introduced in Uzbekistan, leading to notable achievements in the use of innovative, resource-saving technologies [3, 4].

2 Materials and Methods

In recent years, various environmental problems (e.g., unusual weather conditions, dust storms, and a lack of water) have been observed in Uzbekistan. Events such as record-high temperatures, a decrease in cold days, an increase in hot days, and frequent dust storms highlight the severity of the environmental issues [5, 6].

Due to global warming, the temperature has risen by one degree above the norm. In Uzbekistan, it has risen by two degrees.

To improve the ecological situation in Tashkent [1, 7, 8], it is necessary to plant trees in the right place and in the right form [9, 10]. Almost every large city has high levels of pollutants in its atmosphere, which negatively affect the ecosystem and public health.

The environmental performance index compiled by El University scientists evaluated 180 countries in terms of environmental conditions and ecosystem stability in 11 categories of performance indicators. According to research and air quality indicators, Uzbekistan ranks 177th among 180 countries worldwide.

This is not surprising for the residents of Uzbekistan, especially Tashkent. According to the IQAir platform [11], the air quality in Tashkent is very bad. In the environmental performance index of El University, the air quality in Uzbekistan is rated as very low. The countries of Nepal, India, and Pakistan are in the next places.

According to this research, air quality in Uzbekistan is about twice as bad as in Zimbabwe and Kazakhstan, and it is three times worse than in Algeria and Iran. Tashkent is becoming one of the cities with the most polluted air. The area of green areas is currently 7000 ha. Its expansion to 25,000 ha. is part of the city's Master Plan [9] until 2045. In this case, the creation of greening systems stems from the need to improve the urban microclimate. Tashkent ranks 82nd in terms of air quality and pollution.

There are many apparent causes of air pollution: industrial and household emissions from burning coal, wood, or diesel fuel [12, 13], exhaust gases from cars, etc. In Uzbekistan, the list of causes includes the destruction of green spaces and parks for urban development, poor public transportation management, the unlimited expansion of road networks, and the policy promoting mass automobile ownership.

The number of premature deaths due to air pollution was 8.8 million in 2019. Air pollution is mostly caused by human activities [10], including industrial enterprises, transport, agriculture, and household waste. The unique characteristics of the natural environment also play a significant role.

3 Results

When emergencies occur, information about the real situation makes it possible to eliminate the consequences [10] and organize effective measures to reduce their impact on the population and the environment.

The level of automobile ownership in cities is increasing year by year. The problem of insufficient parking space in large cities, where the number of private vehicles continues to rise, is becoming increasingly urgent. Considering the annual growth of the vehicle fleet, this issue needs to be addressed promptly. According to the statistics agency, as of January 1, 2023, the number of cars owned by individuals in the Republic of Uzbekistan [14] was 3.4 million.

For many years, parking lots have been built in Tashkent using asphalt and waterproof concrete [7]. Such parking lots cause environmental problems such as flooding, water pollution, and the urban heat island effect. The maximum heat index in Tashkent is observed in July when the maximum temperature reaches 38 °C and the humidity is 34%.

In turn, along with the growing number of motor vehicles in Tashkent, the demand for parking lots, including roadside parking spaces, is increasing every day. Due to the lack of systemic organization of roadside parking spaces, we can witness an increase in the number of streets in the central streets of Tashkent, where parking spaces take up more space than traffic.

Overheating a vehicle after being parked in direct sunlight increases energy consumption from air conditioning and poses safety risks, such as traffic accidents and heat stroke for passengers [15].

The temperature can rise to 100 °C in a car in direct sunlight.

Eco-parking makes it possible to determine the temperature in the interior of a car parked in direct sunlight. This indicates the rate at which the internal temperature of the vehicle can increase [12, 13].

The authors compared the air temperature inside and outside the car [16] after 60 min of exposure to the sun at 38 °C (Table 1).

Air pollution knows no borders. Emissions from one country can travel and settle in another country, sometimes over distances of thousands of kilometers. The air

Table 1 Air temperature inside and outside the car

Approximate car interior air temperature and elapsed time						
Elapsed time (minutes)	Outside air temperature (°C)					
	21.1	23.9	26.7	29.4	32.2	35
0	21.1	23.9	26.7	29.4	32.2	35
10	31.7	34.4	37.2	40	42.8	45.6
20	37.2	40	42.8	45.6	48.3	51.1
30	40	42.8	45.6	48.3	51.1	53.9
40	42.2	45	47.8	50.6	53.3	56.1
50	43.9	46.7	49.4	52.2	55	57.8
60	45	47.8	50.6	53.3	56.1	58.9
>60	46.1	48.9	51.7	54.4	57.2	60

Source Developed by the authors based on [16]

pollution index (API) of Tashkent city was examined as part of this research [11, 17] (Table 2).

On September 16, 2022, the Council of People's Deputies of the city of Tashkent adopted the decision "On approval of the master plan for greening of the city of Tashkent."

Green Trees are the only natural way to clean the air. During photosynthesis, one hectare of green trees gives off enough oxygen for 200 people to breathe easily. In this process, the plant absorbs up to 8 kg of carbon dioxide every hour. In other

Table 2 Elements of microclimatic efficiency and external improvement of green spaces in the conditions of overheating of the suburbs

Landscaping elements	A decrease in air temperature, °C	Increase in relative humidity, %	Reduce wind speed, %	Reduce the intensity of direct solar radiation, %	Surface lower the temperature, °C
Green fields	3.5–5.5	10–20	50–75	95–100	20–25
A group of trees	1–1.5	4–6	20–40	94–96	12–20
Simple tree planting	1–1.5	4–7	30–50	95	12–19
Lawn, flowers	0.5	1–4	–	–	6–12
Fontan	1.5–3.5	5–10	–	–	–
Water pool for children	–	–	–	–	–
The pergola is covered with plants	1–1.5	–	20–30	80	–

Source Developed by the authors based on [2]

words, 50 square meters of green trees should be available per person to maintain the optimal amount of air in the city. Therefore, this document sets the goal of increasing the capital's greenness level to 30% by 2024 [3].

4 Conclusion

The most important characteristic of the urban environment is the microclimate. Its condition is mainly determined by anthropogenic environmental factors, primarily its pollution.

The concept of the master plan of Tashkent determined that the air temperature in the existing green zones is five degrees lower than in sparse areas and streets.

According to the master plan of Tashkent, there are plans to enhance the banks of the city's five green arteries along the five channels that cross Tashkent, plant saplings, and create a 130-km-long "green ring" along the Tashkent Highway (THAY). The shortage of parking spaces in Tashkent has led to various problems, with cars frequently parked along the edges of city streets and roads. To mitigate these issues and help moderate the city's climate, the construction of ecological parking lots is recommended. These environmentally friendly parking solutions are widely used around the world. Their primary purpose is to protect the ecological environment while ensuring smooth traffic flow. In these parking lots, special lawn fences are installed to preserve the grass cover, preventing its damage and degradation.

Nowadays, the creation of ecological parking spaces in the city of Tashkent is considered the only option for solving two problems at the same time: the organization of parking lots and beautification. Before carrying out work on the organization of an ecological parking lot, it is necessary to determine the territory of the future parking lot. For this purpose, it is necessary to collect, analyze, and determine the work of the geobase based on GIS technologies.

References

1. Sadikov I, Tursunboev F, Djumaev A, Pulatova T (2022) Around roads and recreation areas-reducing environmental damage to vehicles through greening in accordance with environmental landscape. AIP Conf Proc 2432:030080. https://doi.org/10.1063/5.0089600
2. Ministry of Construction and Housing and Communal Services of the Republic of Uzbekistan (2024) Order on approval of urban planning norms and rules SHNK 2.07.01-23 'Urban planning for the development and construction of territories of populated areas' (May 20, 2024 No. 01/2-20). Tashkent, Uzbekistan: LexUZ
3. Coutts AM, White EC, Tapper NJ, Beringer J, Livesley SJ (2016) Temperature and human thermal comfort effects of street trees across three contrasting street canyon environments. Theoret Appl Climatol 124:55–68. https://doi.org/10.1007/s00704-015-1409-y

4. Pegin PA (2011) Impact assessment of the effect of solar glare on the motorway throughput. Vestn Orenburg State Univ 10(129):94–99. Retrieved from https://cyberleninka.ru/article/n/otsenka-vliyaniya-effekta-solnechnogo-oslepleniya-na-propusknuyu-sposobnost-avtomagistrali. Accessed 17 Sept 2024
5. Council of People's Deputies of the city of Tashkent (2002) Decision on approval of the master plan for greening of the city of Tashkent (September 16, 2002 No. VI-70-241-14-0-K/22. Tashkent, Uzbekistan
6. President of the Republic of Uzbekistan (2017) Resolution "On measures to improve the system of landscaping and architectural and landscape design of highways" September 11, 2017 No. PP-3262. Tashkent, Uzbekistan
7. Beketov AK, Saydametova FJ, Ergashova MZ, Khalimova SR (2023) Foreign experience in urban streets management system. Prospects for international specialists in the field of transport: International scientific-practical conference. Tashkent State Transport University, Tashkent, Uzbekistan, pp 528–533
8. Gidro Group (n.d.) Production catalog. Retrieved from https://gidrogroup.ru. Accessed 18 Sept 2024
9. Lv Y, Huang A, Yang J, Xu J, Yang R (2021) Improving cabin thermal environment of parked vehicles under direct sunlight using a daytime radiative cooling cover. Appl Therm Eng 190:116776. https://doi.org/10.1016/j.applthermaleng.2021.116776
10. Magdin K, Mavrin V, Gritsenko A (2021) Reducing the environmental load on urbanized areas by optimizing the parking lots location. IOP Conf Ser Earth Environ Sci 666:052054. https://doi.org/10.1088/1755-1315/666/5/052054
11. IQ Air (2024) Explore your air quality. Retrieved from https://www.iqair.com/us/. Accessed 18 Sept 2024
12. Ali-Toudert F, Mayer H (2007) Effects of asymmetry, galleries, overhanging façades and vegetation on thermal comfort in urban street canyons. Sol Energy 81(6):742–754. https://doi.org/10.1016/j.solener.2006.10.007
13. Bröde P, Fiala D, Błażejczyk K, Holmér I, Jendritzky G, Kampmann B, Havenith G (2012) Deriving the operational procedure for the universal thermal climate index (UTCI). Int J Biometeorol 56:481–494. https://doi.org/10.1007/s00484-011-0454-1
14. Ergashova MZ (2024) Features of greening of city streets and roads in hot climate conditions (in the case of Termiz). Master's thesis. Tashkent, Uzbekistan
15. Sadikov I, Ergashova M (2022) The impact of city streets and small roads on traffic safety. Eurasian J Acad Res 2(13):1367–1376
16. Good Calculators (n.d.) Online calculators. Retrieved from https://goodcalculators.com/. Accessed 18 Sept 2024
17. Statistics Agency under the President of the Republic of Uzbekistan (2024) Statistical data. Retrieved from https://stat.uz/uz/. Accessed 18 Sept 2024

An Optimal Method of Determining the Railway Track Plan

Abdullaaziz Artykbaev, Mokhiniso M. Toshmatova, and Magrurbek Y. Ergashaliyev

Abstract The classical plan of the railway is seen as a flat curve characterized by the projection of the car's center of gravity onto the horizontal plane. The railway plan is described as a single flat curve. The authors argue that the railway plan on the curve will be a spatial curve. Therefore, it is recommended to consider the spatial aspects in curved sections, with each railway curve evaluated individually. In this context, the inner rail's graph on the curve will be a flat circle; the outer graph will form a spatial curve. The distance between the corresponding points of the curves should be equal to the path's width. This research focuses on the mathematical definition of equations for curves, which describe the lines of railway rails. The inner rail is treated as a flat curve, while the outer rail is represented by a spatial curve. The equations of these curves are formulated based on the parameters of the specific route. The primary parameter is the distance between the ends of the straight sections connected by the curved section under study. The spatial curve, representing the outer rail, is chosen so that its projection onto the horizontal plane forms a graph of the clothoid curve.

Keywords Flat curve · Spatial curve · Clothoid · Railway plan · Outer rail · Inner rail · Curvature · Torsion · Radius of the railway plan

JEL codes R41 · R42 · L92 · C61

1 Introduction

Railway track planning is an essential component of infrastructure development, requiring thorough attention to numerous factors to guarantee efficiency and safety. This discussion explores an optimal strategy for devising railway track plans. This approach entails a systemic method designed to enhance functionality, reduce costs, and mitigate potential risks. The process involves detailed analysis and careful

A. Artykbaev · M. M. Toshmatova (✉) · M. Y. Ergashaliyev
Tashkent State Transport University, Tashkent, Uzbekistan
e-mail: toshmatova_mm@mail.ru

S. M. Sultanova et al. (eds.), *Sustainable Development of Transport*, Lecture Notes in Networks and Systems 1366, https://doi.org/10.1007/978-3-031-88846-5_2

coordination, ensuring that the resulting track layout meets the highest operational performance and safety standards.

2 Formulation of the Problem

A railway track consists of two rails when located on a horizontal plane, similar to two parallel straight lines. In Uzbekistan, the distance between rails is 1520 mm. However, in the world, there are roads with a distance between rails of 1340 mm [10, pp. 46–48; 14].

The plane of a railway track is a curve formed by the movement of a point symmetrically located between the rails, which is a projection of the center of gravity of the car moving along the track [1].

The railway track plan consists of straight lines and a curved part. Moreover, the curve often connects straight parts of the road, which form a certain constant angle with each other.

The studies of V. A. Buchkin, V. I. Struchenkov, I. V. Turbin, D. M. Sheidwasser, E. P. Lengenkov, R. B. Schattke, S. S. Tsitko, and others set out methods for drawing up a railway track plan and a complete description of methods for designing a longitudinal profile [1, 9, 13].

The historical plan of the railway track was considered as a flat curve. However, since the beginning of the twenty-first century, there have been works that consider the route plan as a spatial curve [4, 5, 15].

To reconstruct the plan of existing railway tracks, a mathematical model of the railway track route, as a spatial curve, is developed.

In this research, the authors propose defining the railway track plan as two curves depicting the movement of railway rails.

3 Methodology

The difficulty of drawing up a railway track plan is mainly manifested in the curved part of the road. These difficulties are primarily related to the geographical location of the road, environmental requirements, and the technical characteristics of the train's movement.

This research aims to find a mathematical model of railway tracks that provides the most accurate coordinate data on the location of the rails relative to the horizontal plane. Therefore, the authors will try to build an abstract mathematical diagram of the curved part of the railway track. For this purpose, the authors consider two straight parts connecting to the curve to belong to the same horizontal plane and make an angle of α between them. In general, the angle value α can take an arbitrary positive value [11, pp. 48–62].

We might consider $0 < \alpha < \pi$. When both straight parts lie on the same horizontal plane, the angle between the straight parts satisfies this condition.

When drawing up a railway track plan, the main parameter is the radius of the curved part, which is inversely proportional to the curvature of the road.

The radius of the curve must satisfy different technical requirements [6]. Among these requirements, the organic integration from below is crucial, as the composition cannot rotate at small radii. Additionally, there is an upper limit for the radius of the curved section of the railway [7].

The route's geometric pattern significantly depends on the radius of the curved part of the railway plan. To achieve a fairly accurate mathematical abstraction, the authors take the main parameter of the road to be the distance between the ends of the straight parts of the road. This parameter gives us the opportunity to select the best radius of the curved part. Let us translate the technical conditions that the radius of the curved part must satisfy to our parameter (i.e., the distance between the ends of the straight parts).

The curved part does not consist of only one arc of a circle. It has a part at the beginning and end of the curve that allows the wheel to rise and fall when moving along the outer rail.

Moreover, the length of the ascent (descent) should not be less than the length of two wagons, which are 50 m [8].

The distance between the straight parts is the chord of the arc of the curved part. Therefore, its length must be determined so that it meets the technical requirement of the curved part of the road.

Therefore, the authors consider the straighter parts of the road to be given as rays starting from the specified points of the horizontal plane, the distance between the ends of these points, and the angle between these.

To determine the spatial coordinates of the railway route, the authors will use the Cartesian coordinate system. Moreover, the authors consider the curve representing the inner rail as a flat curve and the curve representing the outer rail as a spatial one.

From the technical and economic side, it is considered the most effective when the radius of the curve part is the only one. Therefore, the authors consider the graph of the inner rail to be another one of the circles.

Let l be the distance between the ends of the straight parts that intersect and form an angle of α. The authors consider the values l and α—to be the parameters of the considered problem.

On the horizontal plane, which contains the straight parts of the road, the authors introduce a Cartesian coordinate system.

The authors denote the beginning of the rays as A and B; a and b are images of rectilinear parts. Then $AB = l$. The curve corresponding to the inner part of the road connecting points A and B is denoted as γ_1.

When γ_1 is another circle touching the straight parts a and b at points A and B, respectively, the movement of the train at the transition points will be the smoothest and most continuous. Therefore, it is necessary to choose a circle tangent to the straight parts at points A and B. From the tangent property, the circle's radius at the points of contact will be perpendicular to the straight parts of the road. This means

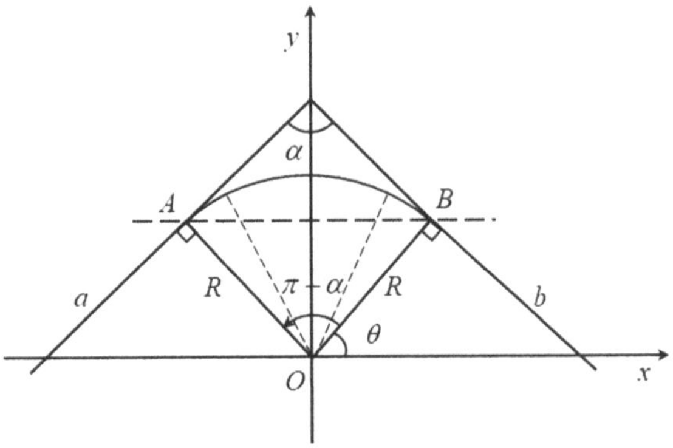

Fig. 1 Scheme of the arc part of the railway track plan, where a and b rays correspond to the straight part; α is the angle between these rays; and the angle between these tangents is the curve between points *A* and *B* of the curved section of the plan. *Source* Developed by the authors

that the center of the circle corresponds to the point of intersection of straight lines perpendicular to straight parts of the road. The authors denote the point of intersection of perpendicular lines as *O* and select it as the origin of the Cartesian coordinate system. Through points *A* and *B*, the authors draw a straight line; the abscissa axis *Ox* is chosen parallel to this straight line. The ordinate axis is perpendicular to the *OX* axis and the positive direction towards the segment *AB*. Moreover, the segment $AB = l$ is a chord of a circle representing the curve γ_1. The axis will be perpendicular to *AB*, dividing it into two equal parts (Fig. 1).

With the chosen Cartesian coordinate system, curve γ_1 has the following parametric equation

$$\begin{cases} x = \frac{l}{2\cos\frac{\alpha}{2}}\cos\theta \\ y = \frac{l}{2\cos\frac{\alpha}{2}}\sin\theta \end{cases} \left(\frac{\alpha}{2} \le \theta \le \pi - \frac{\alpha}{2}\right)$$

In the formula above, $R = \frac{l}{2\cos\frac{\alpha}{2}}$ is the radius of the circle; θ is the angle between the radius and the positive direction of the axis *Ox*.

To resolve the equation of the curve representing the graph of the external part, the authors supplement the coordinate system with an axis perpendicular to the horizontal plane.

To determine the equation of the curve γ_2, which is the graph of the outer part, the authors supplement the coordinate system *Oxy* with axis *Oz* perpendicular to the horizontal plane.

Recall that the outer curve γ_2 is spatial. We assume that the height of the curve relative to the horizontal plane is parallel to the axis *Oz*. We introduce the following notation. The intersection point of radii *OA* and *OB* with the second branch of the straight part of the road is designated *C* and *D*, respectively. Through *h*, we denote the necessary falls of the outer rail at a given speed *V* and the corresponding curve radius γ_2. On the curve γ_2, we mark symmetrical points *E* and *K* so that $\angle EOC = \angle KOD = \beta$. A part of the curve γ_2, enclosed between points *CE* and *DK*, is the

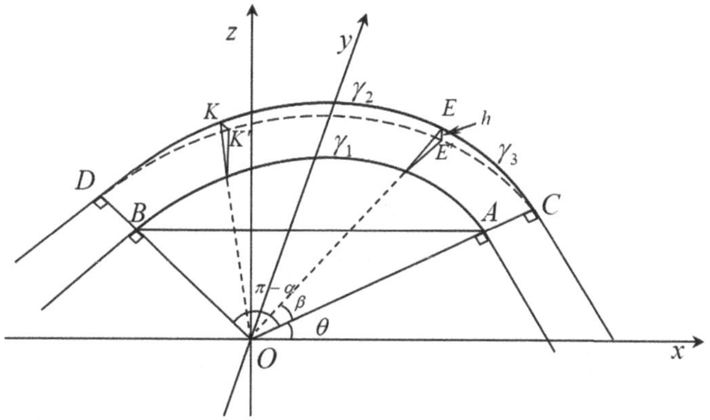

Fig. 2 The distance $AB = l$ between the ends of straight sections of a road. Curve γ_1—graph of the inner rail. The curves on the graph of the outer rail are spatial; γ_3—is the transition part of the graph. *Source* Developed by the authors

curved part where the ramp is made from the horizontal plane to a height of h. The rest of the curve γ_2, enclosed between points EK, has a constant height of h (Fig. 2).

Given that the width of the railway between the tracks is $d = 1520$ mm and the platform is h, we can formulate the equation γ_2, which is enclosed between points EK. The equation for this part of the curve has the following form:

$$\begin{cases} x = \left(\frac{l}{2\cos\frac{\alpha}{2}} + \sqrt{d^2 - h^2}\right)\cos\theta, \\ y = \left(\frac{l}{2\cos\frac{\alpha}{2}} + \sqrt{d^2 - h^2}\right)\sin\theta, \\ z = h. \end{cases}$$

where is $\frac{\alpha}{2} + \beta \leq \theta \leq \pi - \frac{\alpha}{2} + \beta$.

The parameter β is selected according to the requirement for a curved part of the road, in which the length of the curved part should not be less than the length of two cars (i.e., more than 50 m) [12, pp. 63–69].

The sector of the railway track plan in Fig. 2, determined by the parameter β, is the lifting part of the track. In Fig. 2, this part of the outer path is designated by the arc CE and KD. In this part of the outer rail, the car wheel rises from a horizontal position to an inclined one; at point C, the lifting height should reach the value h.

If it denotes the height of the rise through $\tilde{z}$, then it takes on the value at point C, 0 at point E. For convenience, we consider the arc KD to be symmetrical to the arc EC—relative to the plane yOz.

Through E', the authors denote the projection of point E on the plane xOy and consider the curved part CE' to be the projection of curve CE on the horizontal plane.

Usually, when considering the plan of a railway, it is considered flat. The transitional part from the straight part to the curve is considered an arc of the clothoid.

Let us look at some necessary facts from the clothoid. A clothoid, also called a Carnot spiral, is a flat curve in which the radius of curvature varies linearly as a function of the arc length.

This allows the curve to be used as a transition arc in road construction. If the road section in plan has the shape of a clothoid section, then the steering wheel of the car turns smoothly when turning; the centrifugal force increases smoothly (linearly), from zero to a certain value, without lateral jerks. This bend in the road makes it possible to take a turn without significantly reducing the speed. Euler was the first to solve this problem of bending a road. Not knowing about Euler's solution, William Rankine proposed a cubic parabola as a curve with a smooth change in the centrifugal force on turns—a third-order curve, which is approximately a section of a clothoid at not very large angles of rotation.

The clothoid is chosen in the construction of highways. However, it is also used to draw up the plan for railway tracks. When a curved portion of a railroad track plan is treated as a flat curve, we use this curve as a projection of the spatial part [2].

A clothoid is a flat curve in which the radius of curvature varies linearly as a function of the arc length:

$$\begin{cases} x = \int\limits_0^t \cos\left(\frac{\pi u^2}{2}\right) du, \\ y = \int\limits_0^t \sin\left(\frac{\pi u^2}{2}\right) du. \end{cases} \tag{1}$$

The radius of curvature is inversely proportional to the curvature of the curve, following the curvature of the curve CE' can be taken as

$$R = \frac{1}{m \cdot s + n}$$

where m and n are constants; s is the arc length.

The equation of the clothoid coinciding with curve CE' in our choice of Cartesian coordinate system has the following form:

$$x = \left[\left(\frac{l}{2\cos\frac{\alpha}{2}} + \sqrt{d^2 - h^2} \right) \cos\beta - \int\limits_0^t \cos(\frac{\pi u^2}{2}) du \right] \cos\beta,$$

$$y = \left[\left(\frac{l}{2\cos\frac{\alpha}{2}} + \sqrt{d^2 - h^2} \right) \sin\beta - \int\limits_0^t \sin(\frac{\pi u^2}{2}) du \right] \sin\beta$$

Since the curved part CE' must deviate from the straight line, the straight part of the outer rail ends at point C.

By analogy with the classical methods of drawing up a railway track plan, we consider projection CE' to be an arc of the clothoid.

Now, the authors determine the height of the curved part CE. As noted above, we consider it $\tilde{z}$.

4 Results

The equation of the curved part CE that we are looking at is considered to be given by vector equation:

$$r(x) = x(s)i + y(s)\mathrm{j} + \tilde{z}(s)k \tag{2}$$

where $x(s)$ and $y(s)$ are a parametric projection equation, which is the graph of the clothoid given by Eq. (1). In Eq. (2), the unknown function is function $\tilde{z}(s)$.

We choose function $\tilde{z}(s)$ in such a way that the torsion of the curve $\tilde{z}(s)$ changed linearly with respect to the arc length of the curve.

The torsion of the curve given by Eq. (2) is calculated by the following formula:

$$\sigma = \frac{1}{k^2}\begin{vmatrix} \dot{x} & \dot{y} & \dot{\tilde{z}} \\ \ddot{x} & \ddot{y} & \ddot{\tilde{z}} \\ \dddot{x} & \dddot{y} & \dddot{\tilde{z}} \end{vmatrix}$$

From the conditions that the torsion σ of curve CE varies linearly with respect to the length of the arc, we obtain the differential equality:

$$\sigma = \frac{1}{k^2}\begin{vmatrix} \dot{x} & \dot{y} & \dot{\tilde{z}} \\ \ddot{x} & \ddot{y} & \ddot{\tilde{z}} \\ \dddot{x} & \dddot{y} & \dddot{\tilde{z}} \end{vmatrix} = (ps + q)(ms + n)^2$$

In this differential equation, the unknown is the function $\tilde{z}$.

Coefficients p and q are determined from the conditions that the curve starts from C and ends at point E.

All parameters of the curved part of the plan are determined by the technical conditions imposed on the road, which must ensure a smooth transition of the curved part while maintaining the speed of the train [3, pp. 77–79].

5 Conclusion

The curved portion of a railway track plan is a spatial curve. Therefore, in the curvature part of the railway track plan, the graph of the curve representing the rail must be determined separately for the inner and outer rail.

Then, the graph of the inner part will be an arc and the outer part will be spatial.

With the given parameters, the trading parameters are the coal between the straight sections of the road, the width between the ends of the straight parts, and the speed, and mass of the train. It is possible to determine the analytical equation of the graph of the road rails. This makes it possible to determine the coordinates of the road with sufficient accuracy.

References

1. Buchkin VA (2001) Methodology of computer-aided design of reconstruction of the plan and profile of railways. Dissertation of candidate of technical sciences. Moscow, Russia: Russian University of Transport (MIIT)
2. Constantin (2016) The clothoid. PwayBlog. Retrieved from https://pwayblog.com/2016/07/03/the-clothoid/. Accessed 23 Jan 2024
3. Dynnikov IA (2019) Classical differential geometry: lecture course. Department of Mechanics and Mathematics of the Lomonosov Moscow State University, Moscow, Russia
4. Dyunin AK, Protsenko AI (1967) Analytical method for designing the reconstruction of railway tracks in plan: textbook. Novosibirsk Institute of Railway Engineers, Novosibirsk, Russia
5. Gawash T, Jilcha K (2023) Design risk modeling and analysis for railway construction projects. Int J Constr Manag 23(14):2488–2498. https://doi.org/10.1080/15623599.2022.2070344
6. Korzhenevich IP (2007) Mathematical model of the layout of an existing railway track. Sci Transp Prog Bull Dnipropetrovsk Natl Univ Railway Transp 15:50–51
7. Krivchenya IN, Dubrovskaya TA (2019) Application of mathematical modeling methods when designing reconstruction of railways. J Transsib Railway Stud 2(38):109–116
8. Lenchenkova EP (2019) Development of mathematical models of the railway route for plan reconstruction. Dissertation of candidate of technical sciences. Russian University of Transport (MIIT), Moscow, Russia
9. Mamitko AA (2013) Automatic construction of the structure of the railway line plan. In: Transport infrastructure of the Siberian region: materials of the IV All-Russian scientific and practical conference with international Participation: in 2 vols, vol 1. Irkutsk State Transport University, Irkutsk, Russia, pp 516–521
10. Mundrey JS (2009) Railway track engineering. McGraw Hill Education, New York, NY
11. Norberg A (2012) Implementing building information modeling within the railway sector. Master of Science Thesis in the Master's programme Geo and Water Engineering. Chalmers University of Technology, Goteborg, Sweden
12. Ponarin AS (1995) Mathematical models in railway routing. Russia, Moscow
13. Popov NP, Usoltsev AO, Glebov VV (2016) Clotoids and railways. Eur Res 2(13):8–9
14. Smith JA, Doe RB (2020) An analysis of railway track gauges. J Rail Rapid Transit 234(5):456–478
15. Verbilo VA, Kozhedub SS (2018) Fundamentals of designing single-track railways: educational method. Belarusian State University of Transport, Gomel, Belarus

Priority Areas of Public Transport Development in Cities

Maxira N. Usmanova, Matluba A. Xadjimuxametova, and Shokirxo'ja Sh. Utkirov

Abstract The speed of traffic on Uzbekistan's highways is increasing every day. This indicator clearly grows in big cities. Tashkent belongs to the category of large cities, with a population that continues to grow each year. Consequently, the number of vehicles on the city's streets is also increasing. Considering these factors, finding a quick solution to these issues is challenging. To address this, a survey was conducted among city residents, and proposals were developed based on the results to improve the situation. Since ensuring traffic safety requires comprehensive measures, we hope that the aforementioned proposals for traffic organization in large cities will also help address the challenges in Tashkent.

Keywords Highways · Traffic accidents · Vehicle · City streets · Traffic jam

JEL codes L62 · L92 · O18 · R00 · R41 · R42

1 Introduction

The economic changes of the Republic of Uzbekistan are consistently implemented [2, 6, 11, 12, 17]. For example, the automobile manufacturing industry was launched, the road construction sector was accelerated, and trade relations with foreign countries were raised to a high level [1, 4, 7, 8]. This naturally led to an annual increase in traffic on the country's highways [5, 13, 15, 16]. This imposes urgent tasks on the experts in the field of traffic safety and the citizens of the country [14].

Tashkent is one of Central Asia's largest and oldest cities, recognized as a major industrial and transport hub in the region. According to official statistics, the city's area is 334.8 km^2, and its population is 2.7 million people, with 2 million being

M. N. Usmanova (✉) · M. A. Xadjimuxametova · S. Sh. Utkirov
Tashkent State Transport University, Tashkent, Uzbekistan
e-mail: usmanovamahira11@gmail.com

S. Sh. Utkirov
e-mail: shokirxuja8687@gmail.com

S. M. Sultanova et al. (eds.), *Sustainable Development of Transport*, Lecture Notes in Networks and Systems 1366, https://doi.org/10.1007/978-3-031-88846-5_3

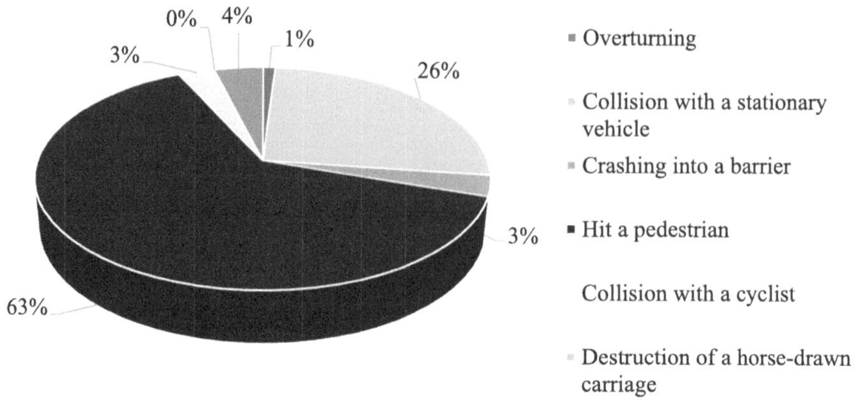

Fig. 1 Cyclogram by types of traffic accidents on the streets of the capital in 2023 [10, 18]. *Source* Developed by the authors based on

mobile. When considering an additional 1.4 million people who are part of the socially active and mobile population, the city's total number of mobile individuals reaches 3.4 million.

2 Methodology

In Tashkent, there is approximately one car for every six citizens. Additionally, over 165,000 motor vehicles enter and leave the city from outside each day [10]. This naturally causes traffic accidents on the city streets (Fig. 1).

According to the cyclogram, in 2023, the most committed type of road traffic accident was running over pedestrians (63%). Ensuring the safety of city residents in transport mobility is the content of this research [14]. Traffic congestion is a serious problem in Tashkent city.

During the research carried out by the experts [3], a social survey was organized to study the attitude of the residents of Tashkent to traffic congestion and their suggestions for reducing it.

3 Results

A total of 891 people participated in the survey. Of them, 60.8% were car drivers. Most respondents are men (89.7%), while 10.3% are women.

Factors causing traffic jams are presented in Table 1.

Based on the survey, the authors obtained the answers showing the indicators of the use of transport other than a private car (Table 2).

Table 1 Factors causing traffic jams

Question about the main cause of the problem (100 people)		
	Reasons	%
	Accidents	9.3
	Other reasons	0.8
	For employees	1.9
	No rules	0.3
	Bad road	31.2
	Weather conditions	0.1
	Traffic light time interval	0.6
	Disorganized drivers are to blame for parking	11.4
Car (passenger)	Irregular passenger traffic	9.9
Public transport (passengers)	Disorder among bus passengers	29.8

Source Developed by the authors

Table 2 Indicators of the use of transport other than a private car

Indicators of the use of transport other than a private car	
Bus	41.8
Subway	27.5
Taxi	24.1
Minibus	1.7
On foot	4.9

Source Developed by the authors

Thus, participants who stated that they do not drive a car chose the following alternative types of transportation:

- 41.8% prefer buses;
- 27.5% selected metro;
- 24.1%—taxi;
- 1.7% use the minibus service;
- 4.9% prefer walking.

The indicators of public transport use are presented in Fig. 2.

In the survey, 74% of participants identified traffic congestion in the capital's streets as a major problem, 21% considered it a moderate issue, and 5% viewed it as less significant compared to other problems. Bus and minibus passengers are particularly affected by traffic jams, with 82.9% of bus users and 83.3% percent of minibus passengers citing congestion as a primary concern in the capital. Among the respondents, 62.2% pointed to increased driver activity due to traffic violations and improper parking as key contributors to congestion. However, 57.6% blamed road defects, 46.8% mentioned commuters, and 41.2% highlighted pedestrian issues.

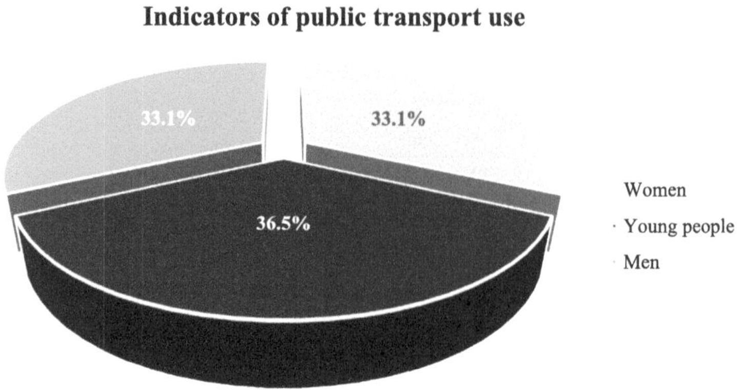

Fig. 2 Indicators of public transport use. *Source* Developed by the authors based on [9]

Additionally, 15.3% of respondents indicated that intersections on the small ring road should be prioritized for regulation, and 12% pointed to intersections on Bunyodkor Avenue. Due to traffic congestion, 41.8% of participants said they allocate more time for their commute, 39.1% choose routes with less traffic, and 16.9% expressed hope for a decrease in traffic.

4 Conclusion

As part of the public transport system reform, Tashkent acquired 700 natural gas-powered buses and 300 electric buses. Earlier this year, 110 buses and 150 electric buses were imported from China. The construction and operation of the Tashkent public highway, spanning 54.8 km and featuring 35 stations, has greatly improved convenience for the city's residents.

Currently, traffic management on city streets is facilitated by advanced organizational and technical tools based on scientific research. Given the ongoing changes in Tashkent's transport and infrastructure, the following proposals are worth considering for improving traffic organization:

- Zoning the city: It is recommended to divide Tashkent into specific areas, such as historical, cultural, educational, production, or high-activity zones.;
- Densely populated areas: Introducing traffic restrictions in heavily populated areas, following best international practices;
- Congestion charges: Implementing a fee system for private vehicles entering the city center;
- Enhancing public transport safety: Establishing dedicated lanes for public passenger transport, and integrating a unified fare collection system within an automated toll payment system.

References

1. Abdulzyanov AR (2024) Sociology of a region. Russ J Reg Stud 4. Retrieved from http://regionsar.ru. Accessed 24 Feb 2024
2. Alavi SS, Mohammadi MR, Soori H, Ghanizadeh M (2017) The Cognitive and psychological factors (personality, driving behavior, and mental illnesses) as predictors in traffic violations. Iran J Psychiatry 12(2):78–86
3. Haghighi MRR, Sayari M, Ghahramani S, Lankarani KB (2020) Social, economic, and legislative factors and global road traffic fatalities. BMC Public Health 20(1):1413. https://doi.org/10.1186/s12889-020-09491-x
4. Heydari S, Hickford A, McIlroy R, Turner J, Bachani AM (2019) Road safety in low-income countries: state of knowledge and future directions. Sustainability 11:6249. https://doi.org/10.3390/su11226249
5. Li X, Wu L, Yang X (2018) Exploring the impact of social economic variables on traffic safety performance in Hong Kong: a time series analysis. Saf Sci 109:67–75. https://doi.org/10.1016/j.ssci.2018.05.010
6. Payani S, Hamid H, Law TH (2019) A review on impact of human factors on road safety with special focus on hazard perception and risk-taking among young drivers. IOP Conf Ser Earth Environ Sci 357:012041. https://doi.org/10.1088/1755-1315/357/1/012041
7. Petridou E, Moustaki M (2000) Human factors in the causation of road traffic crashes. Eur J Epidemiol 16:819–826. https://doi.org/10.1023/a:1007649804201
8. Slesarchuk SV (2019) Analysis of the causes of road accidents involving young drivers. Young Sci 20(258):360–361. Retrieved from https://moluch.ru/archive/258/59169/. Accessed 24 Feb 2024
9. Sun L-L, Liu D, Chen T, He M-T (2019) Road traffic safety: an analysis of the cross-effects of economic, road and population factors. Chin J Traumatol 22(5):290–295. https://doi.org/10.1016/j.cjtee.2019.07.004
10. Usmanova MN (1991) Mathematical and statistical modeling of road safety processes (on the materials of the Minavtotrans of the UzSSR). Dissertation of Candidate of Economic Sciences. Tashkent, Uzbek SSR
11. Usmanova MN (2016) An approach to assessing factors of driver skills when ensuring traffic safety. TADI Bull: Sci Tech J 4:90–93
12. Usmanova MN (2017) Determination of priority areas of activity of the road safety service at motor transport enterprises. TADI Bull Sci Tech J 3–4:102–106
13. Usmanova MN (2019) Directions of traffic organization in big cities. Bus Expert 8:42–44
14. Usmanova MN, Rizaev ShSh (2021) Formation of socio-economic factors affecting traffic safety. Am J Interdisc Innov Res 3(4):25–35. https://doi.org/10.37547/tajiir/Volume03Issue04-05
15. Usmanova MN, Utkirov S, Ogli S (2023) Prevention of traffic incidents involving pedestrians in the city of Tashkent. Am J Lang Literacy Learn STEM Educ 1–9:2993–2769
16. Usmanova MN, Utkirov S, Ogli S (2023) Role of drivers in cause of traffic accidents. Int J Formal Educ 2(11):122–127. Retrieved from https://journals.academiczone.net/index.php/ijfe/article/view/1446. Accessed 24 Feb 2024
17. Whitlock G, Norton R, Clark T, Pledger M, Jackson R, MacMahon S (2003) Motor vehicle driver injury and socioeconomic status: a cohort study with prospective and retrospective driver injuries. J Epidemiol Commun Health 57(7):512–516. https://doi.org/10.1136/jech.57.7.512
18. World Health Organization (2015) Global status report on road safety 2015. World Health Organization, Geneva, Switzerland. Retrieved from https://www.afro.who.int/sites/default/files/2017-06/9789241565066_eng.pdf. Accessed 24 February 2024

The First Metropolitan Subway System: Origins, Design, and Initial Safety

Salaxiddin I. Ismailov, Guldora S. Mustaeva, Turgun S. Ismailov, and Gulchekhra B. Ataeva

Abstract The development of underground subway systems has revolutionized urban transportation, providing efficient and sustainable solutions to address the challenges of congestion and mobility in rapidly expanding cities. This research explores the origins, design, and initial safety considerations of the first metropolitan subway system, focusing on the Paris Metro as a pioneering example. Additionally, the research presents two tables that provide a comprehensive overview of the major milestones in the history of underground subway systems and the current state of selected subway networks worldwide. These tables highlight the global spread of subway systems, their impact on urban transportation, and their continuous evolution to meet the growing demands of contemporary cities. The historical milestones demonstrate the visionary ideas of inventors and city planners. The current status of subway systems showcases their scale, significance, and role in shaping urban mobility. The presented information underscores the importance of underground subway systems as vital components of sustainable urban development and emphasizes their ongoing contributions to improving the quality of life for millions of people worldwide.

Keywords Underground subway systems · Major milestones · Global spread · Urban transportation · Sustainability · Subway networks · Urban mobility

JEL code 038

S. I. Ismailov
Samarkand State Institute of Foreign Languages, Samarkand, Uzbekistan

G. S. Mustaeva · G. B. Ataeva (✉)
Tashkent State Transport University, Tashkent, Uzbekistan
e-mail: gulchexra.ataeva@gmail.com

T. S. Ismailov
Uzbekistan State World Languages University, Tashkent, Uzbekistan

S. M. Sultanova et al. (eds.), *Sustainable Development of Transport*, Lecture Notes in Networks and Systems 1366, https://doi.org/10.1007/978-3-031-88846-5_4

1 Introduction

The concept of a subterranean transportation system dates back to ancient times, with mentions in Greek mythology and Roman engineering feats. However, the modern metro or subway system as we know it today has its roots in the nineteenth century when innovators and city planners sought to address the growing transportation needs of rapidly expanding urban centers [12].

The first subway system was proposed for London by Charles Pearson, a city solicitor, as part of a city-improvement plan shortly after the opening of the Thames Tunnel in 1843. After ten years of discussion, the Parliament authorized the construction of 3.75 miles (6 km) of underground railway between Farringdon Street and Bishop's Road, Paddington [1]. Work on the Metropolitan Railway began in 1860 by cut-and-cover methods (i.e., by making trenches along the streets, giving them brick sides, providing girders or a brick arch for the roof, and restoring the roadway on top) [4]. On January 10, 1863, the line was opened using steam locomotives that burned coke and, later, coal. Despite sulfurous fumes released by the locomotive, the line was a success from its opening, carrying 9.5 million passengers in the first year of its existence. In 1866, the City of London and Southwark Subway Company (later the City and South London Railway) began work on their "tube" line, using a tunneling shield developed by Greathead [5].

Charles Tyson Yerkes from America and Émile Roux from Paris played crucial roles in bringing modern subway systems to life in the late nineteenth century. The design and construction of the world's first metro system, the Paris Métro (Métropolitain), began in the late 1890s. The Parisian government aimed to create an underground train network to reduce street congestion and offer efficient transportation for the city's growing population.

The first metro cars were designed with safety in mind, featuring low center-of-gravity designs and robust construction to minimize the risk of derailment and collisions. The tracks and tunnels were built using cutting-edge engineering techniques, ensuring stability and durability. However, at the time, the level of safety we expect from modern metros was not yet achieved. There were still accidents and fatalities in the early years of operation due to the lack of experience and advanced technology.

The construction of the Paris Metro took several years, with the first line (Line 1) opening on July 19, 1900 [9]. The project faced various challenges, including financial constraints, labor disputes, and technical difficulties. Initial estimates put the cost of constructing the first line at around 350 million francs, which equates to approximately $62 million in today's currency.

The construction process involved digging tunnels through dense urban areas using a combination of hand-mining techniques and specialized machinery. The workforce included thousands of laborers who worked tirelessly in often cramped and hazardous conditions. Despite these challenges, the project was completed within an estimated timeframe of 8–10 years [9].

The first modern metro system, the Paris Metro, emerged from the visionary ideas and efforts of inventors like Charles Tyson Yerkes and Émile Roux, as well as the dedication of thousands of workers who built a safe and efficient means of transportation for the people of Paris. Although early subway systems did not meet the stringent safety standards of today, they laid the foundation for future advancements in underground transportation that continue to shape our urban landscapes.

2 Methodology

The data for this research was gathered from a variety of sources, including official metro system websites, urban transportation reports, and academic journals. Specific focus was placed on obtaining the most current data available as of 2023. Data points included the number of stations, number of lines, daily passenger flow, and operational hours for each metro system. Additionally, historical milestones were documented to provide context to the evolution of each system.

The cities selected for this research represent a diverse range of geographical locations and urban densities. The inclusion criteria were based on the historical significance, scale, and operational efficiency of the metro systems. Well-established and emerging metro systems were included to provide a comprehensive overview of global trends.

The collected data was systemically organized into tables for comparative analysis. Table 1 outlines major milestones in the history of underground subway systems, detailing the year of inception and the initial number of stations. Table 2 provides an in-depth look at the current state of selected metro systems, including the number of stations, lines, daily passenger flow, and working hours.

Descriptive statistics were employed to analyze the data, focusing on mean values, ranges, and standard deviations to identify patterns and outliers. Correlation analysis was conducted to explore relationships between the number of stations, lines, and daily passenger flow.

3 Results

The term "metro" originated from the development of the Paris Metro system in the late nineteenth century. The company responsible for constructing and operating the system was the Compagnie du Chemin de Fer Métropolitain de Paris, which translates to the Metropolitan Railway Company of Paris.

As the construction of the Paris Metro progressed, people started referring to the new transportation system by a shortened version of the company's name, calling it "Le Métropolitain." Over time, this name was further simplified to "Métro." The term quickly gained popularity and became the common way to refer to underground rapid transit systems in France and around the world.

Table 1 Major milestones in the history of underground subway systems

No	City	Year	Description	Stations
1	London	1863	The first underground subway system in the world, known as the London Underground or the Tube, was opened	9
2	New York City	1868	The first subway system in the United States, initially called the "New York and Harlem Railroad," opened	28
3	Paris	1900	The Paris Metro was opened for the World's Fair	10
4	Berlin	1902	The Berlin U-Bahn was opened to alleviate traffic congestion	12
5	Tokyo	1927	The Tokyo Subway was opened to accommodate the growing population	37
6	Moscow	1935	The Moscow Metro was opened to showcase the grandeur of the Soviet Union	13
7	Tashkent	1977	The Tashkent Metro in Uzbekistan was opened to accommodate the growing population	9
8	Beijing	1969	The Beijing Subway was opened to accommodate the growing population	17
9	Shanghai	1993	The Shanghai Metro was opened to alleviate traffic congestion	11
10	Seoul	1974	The Seoul Subway was opened to accommodate the growing population	23
11	Mumbai	2014	The Mumbai Metro was opened to alleviate traffic congestion	12

Source Compiled by the authors based on [2, 6, 7, 10, 11, 13]

The influence of the Paris Metro on the spread of the term "metro" is significant. The success and innovation of the Parisian system captured the attention of other cities seeking to address their urban transportation challenges. As cities in different countries began developing underground rail networks, they adopted the term "metro" to describe these systems, following the trend set by the Parisians [3].

Nowadays, the term "metro" is widely recognized and used internationally to refer to underground rapid transit systems in various cities worldwide. It stands as a testament to the pioneering efforts of the Compagnie du Chemin de Fer Métropolitain de Paris and the visionary development of the Paris Metro as a model for urban transportation.

Table 1 provides a summary of major milestones in the history of underground subway systems. It includes the city, the year of opening, a brief description of each subway system, and the number of stations when they were initially opened. The number of stations listed represents the initial count when each subway system was first opened.

The significance of the presented tables goes beyond mere statistics. They provide valuable insights into the evolution of the underground subway systems and their transformative impact on urban transportation. Thus, Table 1 shows that the development of underground metros has been a global phenomenon, with major cities across different continents embracing this mode of transportation.

Table 2 The evolution of the underground system

Country/city	Year opened	Number of stations	Number of lines	Daily passenger flow	Working hours
London, UK	1863	272	11	4.8 million	5 AM–12:30 AM
New York City, USA	1904	472	24	5.6 million	24/7
Paris, France	1900	300	16	5.5 million	5:30 AM–12:30 AM
Tokyo, Japan	1927	304	9	8.7 million	5 AM–12:30 AM
Beijing, China	1969	378	23	10.5 million	5 AM–11:30 PM
Shanghai, China	1993	511	14	11.8 million	5:30 AM–11:30 PM
Moscow, Russia	1935	295	12	6.7 million	5:30 AM–12:30 AM
Seoul, South Korea	1974	231	8	8.3 million	5:30 AM–12:30 AM
Tashkent, Uzbekistan	1977	48	4	600,000	5:30 AM–11:30 PM
Bangkok, Thailand	1999	47	2	700,000	5:30 AM–11:30 PM
Mumbai, India	2014	120	3	4.7 million	5:30 AM–11:30 PM

Source Compiled by the authors based on [2, 6, 7, 10, 11, 13]

The milestones in Table 1 represent pivotal moments in the history of underground subway systems. They mark the turning points when cities recognized the need for efficient and sustainable mass transit solutions. These milestones demonstrate the forward-thinking vision of city planners and engineers who sought to alleviate traffic congestion, enhance connectivity, and improve the quality of life for their citizens.

For example, the opening of the London Underground in 1863 marked the birth of the world's first underground subway system. This groundbreaking achievement paved the way for other cities to follow suit. New York City's subway system, which opened in 1904, became a symbol of progress and urban development in the United States. Paris, Berlin, Tokyo, Moscow, and other major cities quickly followed, constructing their own subway networks to meet the growing demands of their populations.

Table 1 also sheds light on the increasing complexity and scale of these underground systems over time. As cities expanded and populations grew, so did the number of stations and lines. This expansion reflects the ongoing efforts to extend the reach of public transportation, connect new neighborhoods, and accommodate the rising number of commuters. The continuous development of subway systems demonstrates the commitment of cities to offer efficient and accessible transportation options that reduce dependence on private vehicles and contribute to sustainable urban development [8]. Table 2 describes underground metro railways, along with the number of stations, lines, and daily passenger flow. The information is current as of 2023 and is subject to change. Moreover, the daily passenger flow is an approximate number and may vary depending on various factors such as holidays, events, and seasonality. The working hours may also vary during holidays or special events.

Table 2 illustrates the current state of underground subway systems. These systems now serve as lifelines for millions of people, facilitating their daily commutes and enabling seamless travel within cities. The number of stations, lines, and daily passenger flow provides a glimpse into the magnitude of these transportation networks and the significant role they play in shaping urban mobility.

The data in Table 2, while current as of 2023, may have changed due to ongoing expansions and upgrades in various metro systems. As cities continue to evolve, so do their transportation infrastructure. New lines are being added, existing stations are being upgraded, and technological advancements are being implemented to enhance safety, efficiency, and passenger experience.

The presented tables highlight the remarkable evolution of underground subway systems from their inception to their present-day form. They underscore the global impact of these transportation networks on urban development and mobility. As cities continue to grow and face new challenges, underground metros will remain indispensable tools for efficient, sustainable, and accessible transportation. The history and continued evolution of subway systems serve as a testament to human ingenuity, urban planning, and the pursuit of better transportation solutions for our ever-expanding cities.

The comparative analysis highlighted the extensive scale and operational efficiency of metro systems in Beijing and Shanghai, which have the highest daily passenger flows among the systems studied. Tokyo and Seoul also demonstrated significant passenger handling capacities, reflecting their dense urban populations and extensive metro networks.

This research underscores the pivotal role of metro systems in urban transportation, emphasizing their capacity to support large populations and alleviate traffic congestion. The ongoing expansion and modernization efforts across various cities highlight the continuous evolution of these systems to meet the growing demands of urban mobility.

4 Conclusion

The history and development of underground subway systems serve as a testament to human ingenuity, urban planning, and the pursuit of better transportation solutions for our ever-expanding cities. The tables presented in the research highlight the remarkable evolution of underground subway systems from their inception to their present-day form, underscoring the global impact of these transportation networks on urban development and mobility. As cities continue to grow and face new challenges, underground metros will remain indispensable tools for efficient, sustainable, and accessible transportation. The continuous development of subway systems demonstrates the commitment of cities to offer efficient and accessible transportation options that reduce dependence on private vehicles and contribute to sustainable urban development. The milestones in given data represent pivotal moments in the history of underground subway systems, marking the turning points when cities recognized the need for efficient and sustainable mass transit solutions.

References

1. Abernethy S (2020) M. A. C. Horne, London's district railway: a history of the metropolitan district railway company volume two: twentieth century. J Transp Hist 41(1):126–127. https://doi.org/10.1177/0022526619897895
2. Bystrova OV (2022) How we built the subway: about one story of the project History of factories and plants. Philological sciences. Sci Essays High Educ 5:171–178. https://doi.org/10.20339/phs.5-22.171
3. Cao X, Lin Q (2008) The evolution of worldwide metro systems: a study on their scales and network indexes. Acta Geogr Sin 63(12):1257–1267. https://doi.org/10.11821/xb200812003
4. Cudahy BJ (1988) Under the sidewalks of New York: the story of the greatest subway system in the world. Penguin Adult HC/TR, London, UK
5. Day JR (2008) The story of London's underground, 10th edn. Capital Transport Publishing, London, UK
6. Mayer T, Trevien C (2017) The impact of urban public transportation evidence from the Paris region. J Urban Econ 102:1–21. https://doi.org/10.1016/j.jue.2017.07.003
7. Mustaeva GS, Saidivalieva BS, Ataeva GB (2022) Features of professional vocabulary and its difference from terms. AIP Conf Proc 2432(1):060017. https://doi.org/10.1063/5.0089598
8. Mustaeva G, Ataeva G (2023) The role of business English in modern logistics. In: E3S web of conferences, vol 402, p 08008. https://doi.org/10.1051/e3sconf/202340208008
9. Pei A, Xiao F, Yu S, Li L (2022) Efficiency in the evolution of metro networks. Sci Rep 12:8326. https://doi.org/10.1038/s41598-022-12053-3
10. Shryock HS (1957) The natural history of standard metropolitan areas. Am J Sociol 63(2):163–170. http://www.jstor.org/stable/2773902. Accessed 18 February 2024
11. Subway (2024) In encyclopedia Britannica. Retrieved from https://www.britannica.com/technology/subway. Accessed 19 Jan 2024
12. Tashkent Metro. (n.d.). History of the subway. Retrieved from https://tashmetro.uz/en/history-of-the-subway/. Accessed 3 Feb 2024
13. WorldAtlas (n.d.) The world's oldest metro systems. Retrieved from https://www.worldatlas.com/articles/the-world-s-oldest-metro-systems.html. Accessed 3 Feb 2024

Advancing the Economic Development of Transport Systems Through Modeling

Yulduz Kh. Rakhmonova, Mamura B. Yuldosheva, and Diyora A. Ishankulova

Abstract Transportation systems have significantly evolved over the past century, prompting the development of predictive, analytical, and optimization models. Since the 1950s, the trajectory of transport modeling has been shaped by various philosophies of science, influencing assumptions, methodologies, and outcomes. This research advocates for a contemporary philosophy of science that aligns more effectively with current modeling practices, recognizing them as linguistic activities within the broader framework of transport planning. The paper introduces three alternative perspectives—metaphorical, narrative, and aesthetic—as avenues for reimagining transport models. Despite the shift from deterministic to complex systems-based methodologies, existing models encounter challenges in addressing subjective aspects of human behavior within transportation systems. Employing storytelling, the narrative approach enhances communication by integrating subjective elements into transport planning. Meanwhile, the aesthetic approach, which emphasizes visual representation, underscores the importance of design and aesthetics in shaping stakeholder perceptions. The research concludes by proposing future research directions that incorporate formal philosophical thinking into transport modeling research and practice, highlighting the potential of these models to enhance the economic aspects of transportation systems.

Keywords Transport modelling · Philosophy of science · Linguistic activity · Communication process · Metaphorical approach · Narrative approach

JEL code Q56

Y. Kh. Rakhmonova (✉) · M. B. Yuldosheva · D. A. Ishankulova
Tashkent State Transport University, Tashkent, Uzbekistan
e-mail: yulduzrahmonova91@gmail.com

S. M. Sultanova et al. (eds.), *Sustainable Development of Transport*, Lecture Notes in Networks and Systems 1366, https://doi.org/10.1007/978-3-031-88846-5_5

1 Introduction

Transportation systems have evolved significantly over the past century, necessitating the development of models to predict, analyze, and optimize their performance [16]. Since the 1950s, transport modeling has been shaped by various philosophies of science that have influenced the assumptions, methods, and outcomes of these models. However, it is argued that a new philosophy of science is required to better align with current modeling practices, which are increasingly recognized as linguistic activities within the broader context of transport planning [9]. This paper presents three alternative perspectives for rethinking transport models from metaphorical, narrative, and aesthetic viewpoints while suggesting future research directions that integrate formal philosophical thinking into transport modeling research and practice.

Transport modeling evolved from the deterministic approaches of early transportation engineers to more complex systems-based methodologies [7]. Initially influenced by positivist philosophy, which emphasized objective, empirical observation, and quantitative measurement, transport models were designed to provide accurate predictions based on causal relationships between variables. However, as the complexity of transport systems increased, modelers began adopting post-positivist philosophies that acknowledged the inherent uncertainty in such systems [3].

Despite shifts in philosophical thinking, current transport models still struggle to account for the subjective elements of human behavior and decision-making within transportation systems. This shortcoming can be attributed to an incongruity between the philosophy of science adopted by early transport modelers and the actual practices employed today [8]. By adopting alternative perspectives such as metaphorical, narrative, and aesthetic approaches to transport modeling, we can enhance our understanding of transportation systems and promote inclusive decision-making processes while fostering interdisciplinary collaboration between philosophers, transport planners, and modelers.

2 Methods

The methods used in this research encompass a range of innovative approaches to transportation modeling, each serving a specific purpose in enhancing stakeholder engagement, promoting understanding, and facilitating inclusive decision-making in transport planning.

The metaphorical approach to navigation models employs metaphors to represent complex transportation concepts in a more relatable and understandable manner. By likening transport systems to ecosystems or dynamic systems, planners aim to convey the interconnectedness and dynamic nature of urban mobility. This method is used to engage stakeholders in meaningful discussions about the trade-offs between different transportation modes and their impacts on the overall system.

In the narrative approach to transport models, narratives are utilized to present transportation information as compelling stories that resonate with stakeholders' values, beliefs, and emotions. By narrating the experiences of fictional characters like Jane and Henry, planners illustrate the challenges faced by users of the transportation system and the potential benefits of quality transportation solutions. This method fosters inclusive decision-making by incorporating subjective elements into the storytelling process, promoting empathy and understanding among diverse stakeholders.

An aesthetic approach to transport models focuses on the visual representation of transportation systems and their impacts, recognizing the importance of design and aesthetics in shaping stakeholders' perceptions and preferences. By employing various visualization techniques such as interactive maps, infographics, and animated videos, planners aim to make complex transportation data more accessible and engaging to non-technical audiences. This method enhances stakeholder understanding of transportation issues and promotes informed decision-making by aligning transport models with stakeholders' linguistic characteristics.

Overall, these methods were chosen to overcome the limitations of traditional transportation modeling approaches and provide innovative ways to engage stakeholders, communicate complex information, and promote more sustainable and inclusive transportation solutions.

3 Results

Traditionally, transport planners and modelers would follow an objective, quantitative approach to analyze the existing transportation network and forecast the future demand based on population growth and land-use patterns. They might use conventional tools such as linear programming or cellular automata models to optimize the design of a new metro line, focusing primarily on travel time reduction and cost minimization.

3.1 Transport Models as Linguistic Functions in Transportation Planning

To better understand the potential benefits of these new approaches to transportation modeling, we consider a variety of examples, case studies, and tables illustrating their application in real-world situations.

Table 1 Elements of the "Ecosystem of Urban Mobility"

Species	Role in the Ecosystem	Impact on the System
Cars	Predator (high energy consumption, space occupier)	Increased emissions, traffic congestion
Buses	Herbivore (consumer of roads and infrastructure)	Reduced emissions per passenger, efficient use of road space
Bicycles	Omni-feeder (adaptable to various conditions)	Low energy consumption, minimal impact on the environment
Walking	Scavenger (utilizes leftover resources)	Zero-emission, promotes health and social interaction

Source Compiled by the authors based on [5, 7, 9, 16, 17]

3.1.1 A Metaphorical Approach to Navigation Models

Metaphors are powerful tools for communicating complex ideas and making sense to different audiences [2]. Metaphors are powerful communication tools that shape how we perceive and interpret complex objects [2]. In transportation models, metaphors can be used to express abstract concepts in a more relevant way and make sense of them among stakeholders. For example, transport systems can be conceptualized as "ecosystems" or "complex dynamic systems" that evolve over time, emphasizing their dynamic and unpredictable nature [1].

Let us consider the phrase "the ecology of urban mobility" (Table 1). In this case, transport planners in a medium-sized city aim to improve their existing public transport systems by using ecological metaphors to communicate the complexity of cities within the transport system. The transport network is presented as an ecosystem with different modes of transport (e.g., car, bus, and bicycle) function as interconnected species, each with specific roles and specific effects on the system as a whole.

By using this metaphorical approach, transport planners can engage stakeholders in a more meaningful dialogue about the complex trade-offs between different modes of transportation. This communication strategy allows for a better understanding of how seemingly minor changes to one part of the ecosystem (e.g., adding bike lanes) could have significant ripple effects on other elements of the system, ultimately improving the overall health and sustainability of the urban mobility ecosystem.

3.2 Narrative Approach to Transport Models

Narrative travel models focus on storytelling to deliver complex information in an engaging and accessible way [6]. This approach allows planners to present transportation information as a "narrative" consistent with stakeholders' values, beliefs, and emotions. By incorporating subjective elements into the story, planners can promote more inclusive decision-making processes. Narratives provide a way to organize information into coherent narratives that convey meaning and create shared meaning among participants [6]. Drawing on the paradigm of narrative navigation, narratives of navigation can be presented as a story that includes members, opponents, and narratives. This approach incorporates subjective elements such as values, beliefs, and emotions into the travel planning process, increasing communication between stakeholders [10].

Let us take "Jane and Henry" as an example. In this case study, transportation planners in a rapidly growing city innovate public transport systems by presenting transportation information as a story about two protagonists:

- A young mother named Jane has to travel from her suburban home to her job in the city every day;
- Henry, an elderly gentleman, needs regular visits to health facilities [10].

The research highlights the challenges that Jane and Henry face due to the shortcomings of the existing transportation system (e.g., lack of reliability, safety concerns, and long travel times). Additionally, the advantages of quality in the rein such as increased security, decreased environmental impact, and increased social equality.

Case study: Transportation Options in the Story of Jane and Henry.

Scenario 1: Status Quo.

Plotline Summary: Jane faces long commutes, unreliable service, and safety concerns; Henry has difficulty accessing healthcare.

Benefits for Jane and Henry: Neither Jane nor Henry's needs are fully met.

Scenario 2: Invest in Public Transit (Bus).

Plotline Summary: Jane takes an efficient and reliable bus service that reduces her commute time and improves safety; Henry enjoys better access to healthcare facilities via improved bus routes.

Benefits for Jane and Henry: Reduced travel times for Jane and Henry.

Scenario 3: Promote Biking and Walking.

Plotline Summary: Jane bikes or walks to work, enjoying a healthier lifestyle while reducing her carbon footprint; Henry can walk to the nearest healthcare facility with ease.

Benefits for Jane and Henry: Reduced air pollution due to fewer cars on the road. Improved health for both Jane and Henry.

Scenario 4: Develop an Integrated Transport System (Combining Bus, Bike-Sharing, and Walking).

Plotline Summary: Jane takes a bus partway, transfers to bike-sharing for the final leg of her commute, and walks from the station to her workplace; Henry uses public transit and walking to access healthcare facilities.

Benefits for Jane and Henry: The benefit combines the benefits of different modes, offering more flexible travel options for Jane and Henry. Moreover, it reduces air pollution and traffic congestion due to a mix of transportation options and enhances social equity because public transit becomes more accessible and convenient.

By presenting transportation data as a multi-stage narrative, planners can engage stakeholders in meaningful discussions about the potential benefits and trade-offs of different transportation solutions. This approach makes it possible to better understand how the changes made (e.g., bicycle lane design) can have significant impacts on other system elements, ultimately affecting the overall health and its sustainability improved.

"Incorporating the narrative transport model, as demonstrated in the case study of Transportation Options in the Story of Jane and Henry, can enhance our understanding of transport policy and promote inclusive decision-making processes. By aligning transport models with the linguistic characteristics of stakeholders within the broader transport system, this approach facilitates effective communication and engagement." This example illustrates how presenting transportation data as a "narrative" with other information can lead to effective and sustainable solutions to urban transportation challenges.

3.3 Aesthetic Approach to Transport Models

An aesthetic approach to transport modeling focuses on the visual representation of transportation systems and their impacts, emphasizing the importance of design and aesthetics in shaping stakeholders' perceptions and preferences [15]. This approach can be particularly effective when communicating complex information to non-technical audiences. The aesthetic approach to transport modeling focuses on the visual presentation and interpretation of data, emphasizing its artistic and creative aspects [15]. In this context, transport models can be designed to appeal to the senses and emotions of stakeholders, fostering engagement and collaboration in the decision-making process. For instance, the use of colorful infographics or interactive visualizations can make complex information more accessible and appealing to a broader audience [11].

Let us consider the "Transportation Impact Visualization" as an example. In this case study, transport planners in a coastal city aim to reduce the environmental impact of transportation by promoting sustainable modes such as walking, cycling, and public transit [12]. They create an interactive visualization tool that allows users to explore the potential impact of different transportation scenarios on air quality, noise pollution, and traffic congestion.

To elucidate the aesthetic approach to transport modeling, let us consider an example table that showcases different visualization techniques used to represent transportation impact and their potential benefits (Table 2) [4].

Thus, adopting an aesthetic approach to transport modeling involves using various visualization techniques that present transportation data and scenarios in a visually

Table 2 Visualization techniques for an aesthetic approach to transport models

Visualization Type	Description	Benefits for stakeholders
1. Interactive maps	Digital maps that allow users to explore transportation scenarios and their impact on the environment, traffic congestion, and air quality	It enhances understanding of complex data through visual representation
2. Infographics	Graphical representations of data that convey information about transportation systems in an easily digestible and visually appealing manner	It facilitates informed decision-making by stakeholders; improves accessibility to complex information for non-technical audiences; and encourages dialogue among stakeholders by presenting data in a compelling way
3. Animated videos	Short videos that illustrate transportation scenarios and the impact of different transport solutions using animations, simulations, or real-world footage	It engages stakeholders emotionally by telling a story through visuals and sound; demonstrates potential impacts in an easily understandable format; and facilitates shared understanding among diverse audiences
4. Physical models	Three-dimensional models that represent transportation systems or specific transport infrastructure (e.g., bridges, tunnels, or interchanges)	It allows for tangible exploration of different transport scenarios; enhances stakeholder engagement through hands-on interaction with the model; and facilitates understanding of spatial relationships and impacts within a transportation system
5. Virtual Reality (VR)	Immersive, three-dimensional environments that allow users to experience different transportation scenarios in a realistic and interactive manner	It provides an immersive and engaging way for stakeholders to explore transport solutions; enables stakeholders to visualize the impact of their choices on the urban environment; and facilitates informed decision-making by allowing users to experience different scenarios firsthand

Source Compiled by authors based on [1, 9, 10, 13, 14, 16, 18]

appealing and engaging manner. This approach can enhance stakeholder understanding of complex information and promote more inclusive decision-making processes by aligning transport models with their linguistic nature within the broader context of transport planning. Table 2 illustrates how different visualization techniques can be employed to achieve these benefits, ultimately leading to more effective and sustainable solutions for urban mobility challenges.

4 Conclusion

Linguistic philosophy is the base of almost every business as it is commonly used to give customers the necessity of this item by giving and showing case studies. The linguistic philosophy shift proposed in this paper emphasizes the importance of viewing transport modeling as a linguistic activity within the context of communication-based transport planning. By considering alternative perspectives such as metaphorical, narrative, and aesthetic approaches, we can enhance our understanding of transportation systems and promote more inclusive and participatory decision-making processes. This could involve the development of new methodologies that explicitly acknowledge the linguistic nature of transport modeling, as well as the subjective elements of human behavior in transportation systems. Additionally, interdisciplinary collaboration between linguistic philosophy and transport system economy is essential to foster a deeper understanding of the complex relationship between language philosophy, communication, and transport planning.

References

1. Bell D (2019) Intermodal mobility hubs and user needs. Soc Sci 8(2):65. https://doi.org/10.3390/socsci8020065
2. Berg J, Ihlström J (2019) The importance of public transport for mobility and everyday activities among rural residents. Soc Sci 8(2):58. https://doi.org/10.3390/socsci8020058
3. Evangelinos C, Tscharaktschiew S (2021) The valuation of aesthetic preferences and consequences for urban transport Infrastructures. Sustainability 13(9):4977. https://doi.org/10.3390/su13094977
4. Forward SE (2019) Views on public transport and how personal experiences can contribute to a more positive attitude and behavioural change. Soc Sci 8(2):47. https://doi.org/10.3390/socsci8020047
5. Gijsenberg MJ, Verhoef PC (2019) Moving forward: the role of marketing in fostering public transport usage. J Public Policy Mark 38(3):354–371. https://doi.org/10.1177/0743915619846869
6. Gössling S (2023) Extending the theoretical grounding of mobilities research: transport psychology perspectives. Mobilities 18(2):167–183. https://doi.org/10.1080/17450101.2022.2092886
7. Green MC (2021) Transportation into narrative worlds. In Frank LB, Falzone P (eds), Entertainment-education behind the scenes, pp 87–101. Cham, Switzerland: Palgrave Macmillan. https://doi.org/10.1007/978-3-030-63614-2_6
8. Kaszubowski D (2019) A method for the evaluation of urban freight transport models as a tool for improving the delivery of sustainable urban transport policy. Sustainability 11(6):1535. https://doi.org/10.3390/su11061535
9. Loyola M, Shiftan Y, Aviram H, Monterde-i-Bort H (2019) Impact of public transport context situation and culture on mode choice. Soc Sci 8(2):40. https://doi.org/10.3390/socsci8020040
10. Morales M, Guerra J (2021) Quantitative study of the metaphorical projections of Zeit (Time) in Der Zauberberg by Thomas Mann. Open J Modern Linguist 11(5):727–745. https://doi.org/10.4236/ojml.2021.115058
11. Mustaeva GS, Saidivalieva BS, Ataeva GB (2022) Features of professional vocabulary and its difference from terms. AIP Conf Proc 2432(1):060017. https://doi.org/10.1063/5.0089598

12. Mustaeva G, Ataeva G (2023) The role of business English in modern logistics. E3S Web of Conferences, 402, 08008. https://doi.org/10.1051/e3sconf/202340208008
13. Pokharel R, Bertolini L, te Brömmelstroet M (2023) How does transportation facilitate regional economic development? A heuristic mapping of the literature. Transp Res Interdiscip Perspects 19:100817. https://doi.org/10.1016/j.trip.2023.100817
14. Sobral T, Galvão T, Borges J (2019) Visualization of urban mobility data from intelligent transportation systems. Sensors 19(2):332. https://doi.org/10.3390/s19020332
15. Steg L (2007) Sustainable transportation: a psychological perspective. IATSS Res 31(2):58–66. https://doi.org/10.1016/S0386-1112(14)60223-5
16. Stowell F (2021) Power in the 'Organisation': a soft systems perspective. Syst Pract Action Res 34(4):515–535. https://doi.org/10.1007/s11213-020-09541-w
17. Tomaino G, Teow J, Carmon Z, Lee L, Ben-Akiva M, Chen CY, … Zhao J (2020). Mobility as a service (MaaS): The importance of transportation psychology. Market Lett 31:419–428. https://doi.org/10.2139/ssrn.3634537
18. Zhang H, Ke R, Lin Y, Qu D, Li Z (2019) AirInsight: Visual exploration and interpretation of latent patterns and anomalies in air quality data. Sustainability 11(10):2944. https://doi.org/10.3390/su11102944

Statistical Analysis of the Dynamics of Passenger Transportation on Railway Transport of the Republic of Uzbekistan

Gaffor A. Samatov, Khalikul Sh. Chariev, and Akmal S. Kamalov

Abstract The normal distribution plays an important role in constructing interval statistical estimates of the numerical characteristics of random processes. In most cases, under the same conditions, the sum of random variables that do not differ significantly from each other will have a normal distribution based on the central limit theorem of probability theory. In this research, the authors analyzed the average annual number of passengers transported by the railway transport company in the Republic of Uzbekistan. The primary statistical hypothesis assumes a normal distribution. The alternative hypothesis, based on the Shapiro-Wilcoxon test, suggests that the data do not follow a normal distribution. To validate these hypotheses, the authors used Pearson and parametric tests. The authors also calculated the key numerical characteristics of the average annual passenger numbers in the Republic of Uzbekistan. The results of this analysis help reduce uncertainty, enhance the accuracy of passenger flow forecasting, and improve the planning and management practices of the railway transport company in the Republic of Uzbekistan.

Keywords Transport company · Uzbekistan railway · Railway passengers · "Uzbekistan Railways" JSC · Passenger transport

JEL Code L92

1 Introduction

Since the number of passengers (thousands of passengers) transported annually by railway transport of the Republic of Uzbekistan is seasonal, the laws of statistical development of this process can be studied using the theory of time series of

G. A. Samatov (✉) · K. Sh. Chariev
Tashkent State Transport University, Tashkent, Uzbekistan
e-mail: transportlogistikasi@mail.ru

A. S. Kamalov
"Uzbekistan Railways" JSC, Tashkent, Uzbekistan

S. M. Sultanova et al. (eds.), *Sustainable Development of Transport*, Lecture Notes in Networks and Systems 1366, https://doi.org/10.1007/978-3-031-88846-5_6

mathematical statistics. The main challenge lies in identifying guaranteed statistical patterns from the limited amount of data and analysis available and predicting the number of passengers that will be carried in the future with a 90%–95% guarantee.

In the research, the number of passengers transported by "Uzbekistan Railways" JSC in 2013–2022 (thousand passengers) was studied as a discrete dynamic series $\{Y_t, t \in T\}$.

2 Materials and Method

The section of dynamic (time) series, widely used in solving practical problems of mathematical statistics [1–3, 5, 6, 8–12, 16], has been carefully studied and used to solve practical problems in foreign countries. We can indicate the following literature on dynamic lines: Anderson [4], Kendal and Stewart [13], Tikhomirov [15], Sulaimanovet al. [14], Fayziev [7], and others.

According to the [4, 7, 13–15], the number of passengers transported by "Uzbekistan Railways" JSC from 2013 to 2022 (in thousands of passengers) forms a discrete $\{Y_t, t \in T\}$ and stable dynamic series (Table-1): $y = a_0 + a_1 t$ (Fig. 1).

To estimate the unknown parameters in a linear relationship based on statistical data, the least squares method is commonly used:

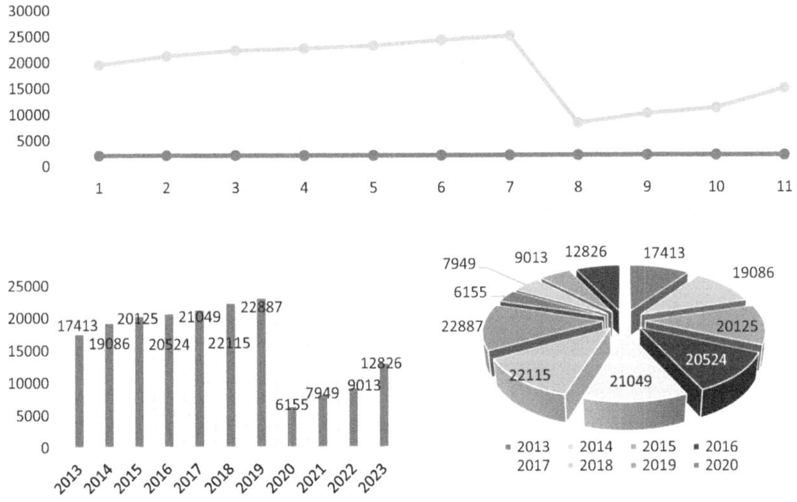

Fig. 1 Trend model characterizing the direction of passenger transportation of "Uzbekistan Railways" JSC. *Source* Developed by the authors

$$\begin{cases} a_0 T + a_1 \sum t = \sum y_t \\ a_0 \sum t + a_1 \sum t^2 = \sum y_t t \end{cases} \tag{1}$$

It is necessary to note that a_0 and a_1 are determined by solving the system of Eq. (1):

$$a_0 = \frac{1}{T} \sum y_t,\ a_1 = \frac{1}{\sum t^2} \sum y_t t. \tag{2}$$

Using the calculations presented in Table 1, the authors determine the unknown parameters involved in the linear relationship (2):

$$\sum y_t = 179141.6 a_0 = \frac{1}{\mathrm{T}} \sum y_t = \frac{179141.6}{11} = 16285.6,$$
$$a_1 = \frac{1}{\Sigma t^2} \sum y_t t = \frac{-126653.13}{110} = -1151.39.$$

Thus, the equation of the trend part, representing the number of passengers transported by "Uzbekistan Railways" JSC for t-years, based on the above calculations, has the following form:

$$y(t) = -1151.39t + 16285.6 \tag{3}$$

Table 1 Analysis of statistical data on the number of passengers transported by "Uzbekistan Railways" JSC

1	2	3	4	5	6	7
S N	Year	y_t– Passengers transported (thousands)	t	t^2	$y_t t$	$y_t t^2$
1	2013	17,413	−5	25	−87,063,13	435,315,65
2	2014	19,086	−4	16	−76,344	305,376
3	2015	20,125	−3	9	−60,375	181,125
4	2016	20,524	−2	4	−41,048	82,096
5	2017	21,049	−1	1	−21,049	21,049
6	2018	22,115	0	0	0	0
7	2019	22,887	1	1	22,887	22,887
8	2020	6155	2	4	12,310	24,620
9	2021	7949	3	9	23,847	71,541
10	2022	9013	4	16	36,052	144,208
11	2023	12,826	5	25	64,130	320,650
Total		179,141.6	0	110	−126,653.13	1,608,867.65

Source Developed by the authors

By placing the value t = 1 in the trend model (3), which represents the main direction of the studied random process, we find out the spread of the infectious disease COVID-19 throughout the world:

$$\Delta Y_t = Y_{t+1} - Y_t,\ \Delta^2 Y_t = Y_{t+1} - Y_t,\ \Delta^3 Y_t = \Delta^2 Y_{t+1} - \Delta^2 Y_t.$$
$$\Delta Y_t = Y_{t+1} - Y_t,\ \Delta^2 Y_t = Y_{t+1} - Y_t,\ \Delta^3 Y_t = \Delta^2 Y_{t+1} - \Delta^2 Y_{t}.,\ \ldots$$

$$V_k = \frac{\sum_{t-k}^{T} \left(\Delta^k y_t\right)^2}{(T-k)C_{2k}^k},\quad k = 1, 2, 3$$

After calculating the values, we see that the coefficients of variation of the finite differences are approximately equal to each other (i.e., $V_1 \approx V_2 \approx V_3$). This is an important observation when studying the characteristics of time series. To further study the time series, the authors calculate the autocorrelation coefficients using the following formula [1–5]:

$$R_L = \frac{\sum_{t=1}^{N-L} Y_t Y_{t+L} - \frac{\sum_{t=1}^{N-L} Y_t \sum_{t=L+1}^{N} Y_t}{N-L}}{\sqrt{\left[\sum_{t=1}^{N-L} Y_t^2 - \frac{\left(\sum_{t=1}^{N-L} Y_t\right)^2}{N-L}\right]\left[\sum_{t=L+1}^{N} Y_t^2 - \frac{\left(\sum_{t=L+1}^{N} Y_t\right)^2}{N-L}\right]}} \tag{4}$$

The calculations of the autocorrelation coefficients R_2, ..., R_5 for the radar data R_L over L years (with L being the lag, i.e., L = 1, 2, 3, 4, 5 years) show that all their values differ from zero. Even if we test this hypothesis using the Durbin-Watson test, it is autocorrelated:

$$d_{kuz} = \sum_{t=1}^{T-1} (Y_{t+1} - Y_t)^2 / \sum_{t=1}^{T} Y_t{}^2.$$

we see that all d_{kuz}-, values found are less than the critical value $d_{krit} = 1.04 d_{kuz} < d_{krit}$. Thus, with 95% confidence, it has been proven that the random process under study exhibits an autocorrelation relationship $y_t = \rho y_{t-1} + \varepsilon_t$, $\rho = \mathrm{Cov}\,(y_t, y_{t+1})$. In other words, the number of passengers transported by the railway transport company in the Republic of Uzbekistan in any given year is part of a complex random process that depends on the number of passengers transported in previous years.

Table 2 Statistical estimates of sample characteristics of the number of passengers transported by rail

Selected characteristics of the number of passengers transported by rail	Statistical estimates of sample characteristics of the number of passengers transported by rail
The average number of passengers transported annually is $\overline{y}_T$- thousand passengers	16,631.3
Sample variance D_T	406,544.0
Sample standard deviation σ_T	637.61
Asymmetry A ς	−0.88
Excess $E_{K\varsigma}$	−1.13
Error of sample mu mean value $\overline{y}_T$	$m_u = \frac{\sigma_u}{\sqrt{n}} = \frac{637.61}{\sqrt{11}} = 192.05$
Limit error m'_u	$m'_u = t\ m_u = 2.23 \cdot 192.05 = 428.27$
Interval statistics (95% li) $\overline{y}_T \pm t\ m_y$ Interval statistical estimation (95% confidence level) $\overline{y}_T \pm t\ m_y$	$\overline{y}_T \pm t\ m_u = 16{,}631.3 \pm 428.27$, (16,203.03; 17,059.57) thousand people
Statistical hypotheses: $H_0 : P(\overline{y}_T < x) = F_{a,\sigma}(x)$ $H_0 : P(\overline{y}_T < x) \neq F_{a,\sigma}(x)$	H_0 is accepted with 95% confidence

Source Developed by the authors

3 Results

Using the ECM 7 × and Excel programs, the authors calculated the significant numerical characteristics of the average number of passengers transported annually by the railway transport company of the Republic of Uzbekistan (Table 2).

Based on this confirmation, an interval statistical assessment of the number of passengers transported was made with a significance level of $\alpha = 0.05$ (i.e., with a 95% guarantee). The estimated figure is (16,203.03; 17,059.57) thousand people.

4 Conclusion

Based on a statistical analysis of the number of passengers transported by "Uzbekistan Railways" JSC, the authors discovered that the passenger numbers follow a complex random process, where the current year's passenger flow depends on the numbers from previous years.

Additionally, the authors calculated key numerical characteristics of the average annual number of passengers transported by the railway company in the Republic of Uzbekistan. The results of this analysis help reduce uncertainty and enhance the accuracy of passenger flow forecasting. In turn, this will improve the planning and management practices of the railway transport company, leading to more efficient operations.

References

1. Abdalimov B, Fayziev AA (2019) Analysis of economic indicators of farming using mathematical modeling. Bulletin Agrarian Sci Uzbekistan 2(76):164–167
2. Abdalimov B, Fayziev AA (2022) Mathematics: textbook. Tashkent, Uzbekistan: Ilm-ziyo-zakovat
3. Achilov MU, Fayziev AA (2019) The analysis of dynamics of fruit and berry productivity grown in Uzbekistan. Int J Res Dev EPRA 4(8):5–9
4. Anderson T (1976) The statistical analysis of time series. Moscow, USSR: MIR. (Original work published 1971)
5. Buriev KhCh, Fayziev AA, Nishanova A (2021) Statistical analysis and forecasting of the dynamics of yield of bakchevyh crops. Bulletin Agrarian Sci Uzbekistan 1(85):47–52
6. Fayziev AA (2021) Application of Markov chain for solving agricultural problems. In:. Farmanov ShK (ed), Proceedings of scientific conference "Modern problems of stochastic analysis", dedicated to the 80th anniversary of Academician, pp 59–62. Tashkent, Uzbekistan: V.I. Romanovskiy Institute of Mathematics of the Academy of Sciences of Uzbekistan
7. Fayziev AA (2022) Mathematical statistics: textbook. Tashkent, Uzbekistan: Ilm-ziyo-zakovat.
8. Fayziev AA, Vakhobov V (2020) Forecasting the dynamics of cotton yield in the Fergana region. J Irrigation Land Reclamation 4(22):68–71
9. Fayziev AA, Yuldashev YX (2024) Statistical analysis and dynamics forecasting pollutants released in the City of Tashkent in the Republic of Uzbekistan. J Res Innov 2(1):49–57. https://doi.org/10.5281/zenodo.10571096
10. Fayziev AA, Farmanov TX, Alladustova FT (2023) Statistical analysis and forecasting of the dynamics of development of cities in the Tashkent region of the Republic of Uzbekistan. J Econ Entrepreneurship 17(10):629–633
11. Fayziev AA, Karabashov OZ, Musaeva NN (2021) Forecasting the dynamics of cotton yield in the Andijan region. Bulletin Sci Educ 8(111):6
12. Fayziev AA, Turgunov A, Mamadaliev Kh, Nasridinov S (2022) Statistical analysis and forecasting of potato yield dynamics in the Republic of Uzbekistan. Proceedings of ICISCT 2022: international conference on information science and communications technologies, pp 1–4. Tashkent, Uzbekistan: Tashkent University of Information Technologies named after Muhammad Al-Khorezmi; IEEE. https://doi.org/10.1109/ICISCT55600.2022.10146743
13. Kendal M, Stewart A (1976) Multivariate statistical analysis and time series. Nauka, Moscow, USSR
14. Sulaimanov BA, Fayziev AA, Fayziev ZhN (2015) Statistical analysis of experimental data. Tashkent, Uzbekistan: ToshDAU
15. Tikhomirov NP (2003) Econometrics. Moscow, Russia: Exam
16. Vagabov V, Fayziev AA (2022) Statistical analysis and forecasting of cotton yield dynamics in Bukhara region. In: 1st International scientific conference "Modern materials science: current issues, achievements and innovations" (ISCMMSIAL-2022). Tashkent, Uzbekistan: Tashkent State University of Transport

Logistics System of Passenger Transportation in Large Cities: Problems and Solutions

Barno O. Kenjaeva and Akmal A. Nazarov

Abstract The research examines the challenges facing passenger transport systems in major cities. The authors use analysis, synthesis, and a systems approach to study problems in the field of logistics systems. The authors identified four main groups of problems: managing peak loads and flows, safety and compliance, environmental issues, and technological limitations and legacy systems. For each of these groups of problems, the authors proposed solutions based on the use of new technologies and tools. As a result of the research, the authors determined the methods for planning and forecasting the dynamics of transport mobility of the population. Moreover, the authors established the importance of considering public opinion when developing regulatory frameworks for planning public passenger transport using logistics methods. The practical application of these research results in the field of public passenger transport logistics undoubtedly yields positive outcomes. In turn, this positively affects the socio-economic development of cities by properly directing investments into urban transport and logistics systems. Therefore, enhancing urban transport systems through logistical methods can be considered a guarantee of sustainable development.

Keywords Passenger · Transport logistics · Large cities · Problems · Solutions · Innovative technologies · Peak loads · Safety · Ecology · Technological limitations

JEL codes L51 · L52 · L90 · L91 · L92 · L98 · O12 · O18 · O21 · O32 · O33 · O38 · P52 · R41 · R42

B. O. Kenjaeva · A. A. Nazarov (✉)
Tashkent State Transport University, Tashkent, Uzbekistan
e-mail: anazarov_07@mail.ru

B. O. Kenjaeva
e-mail: bkenzhaeva@internet.ru

S. M. Sultanova et al. (eds.), *Sustainable Development of Transport*, Lecture Notes in Networks and Systems 1366, https://doi.org/10.1007/978-3-031-88846-5_7

1 Introduction

The logistics system is complex. This system in the field of passenger transportation expands the scope of organization, planning, and management of all passenger flows and logistics management entities. It facilitates the coordination of information and financial flows and the processes involved in transport operations [15].

The research aims to examine the patterns of key characteristics in public passenger transport within the logistics system, which form the core of the study and play a crucial role in assessing their significance in contemporary transport infrastructure. An ineffective logistics system in public passenger transport can result in schedule disruptions, reduced passenger comfort, and longer travel times [4]. This could lead to a reduction in the overall number of passengers using public transport.

Studies show that the data from various transport observations of the population is the only source to fully cover the parameters of the population's demand for transport and the conditions of its satisfaction in the current system [2]. When organizing the work of urban passenger transport, one should consider foreign economic and socio-economic significance [20]. Regularity and reliability play an important role in the optimal organization of public transport [19]. The most important requirements for public passenger transport are their regularity and reliability. This analysis once again shows that a normal working day on a city bus, especially during rush hours, has a significant impact on its performance [1]. Advanced logistics systems are characterized by a high level of development. They make extensive use of advanced technologies (e.g., satellite navigation, automated control systems, etc.) [3].

The use of these methods can ensure high efficiency and reliability of the logistics process. Simultaneously, logistics methods of compliance are used when organizing passenger transportation. As analyses show, logistics methods can be distinguished by peak loads and control flows [13], compliance with security requirements and standards [7], environmental aspects [16], and technological limitations and obsolete systems [12]. This research examines the common problems in public passenger transport across logistics systems in large cities and offers recommended strategies and practical solutions to improve its efficiency. Given that public passenger transport is a key component of the socio-economic indicators of a country's transport system, it holds not only economic innovation potential but also economic attractiveness [23].

It is necessary to address the issue regarding the number of vehicles (intervals between them) and the organization of their operation (based on a schedule or interval), required for servicing the current passenger flow. This issue needs to be resolved for the transition between peak and off-peak times, and vice versa [20].

2 Materials and Methods

The problems of transporting passengers in cities were reflected in the works by Alimukhamedov et al. [2], Andreev [3], Arifdzhanova [4], Arora et al. [5], Butaevet al. [6], Darmogray [7], Farahani et al. [8], Ganiev [9], Gudkov [10], Kandel et al. [11], Kenzhaeva [12], Malyshev [13], Melentev [14], Mironova [15], Mirzaev et al. [16], Morton and Caulfield [17], Nazarov and Ustaboyev [21], Postnikov [22], Shalaik et al. [24], Sorokin [25], Spirin [26], and Zito and Salvo [27].

In public passenger transport, peak loads arise during periods when passenger demand for transport services significantly exceeds supply. This phenomenon can be caused by various factors, such as mass events during holidays, the start of the school year, or peak working days.

Peak loads have several negative consequences for passengers and transport operators. During such peak periods, passengers are often forced to wait at public transport stops for extended periods; traveling in overcrowded buses reduces service comfort and increases inconvenience. Additionally, this situation raises the risk of traffic accidents. Peak times also create challenging conditions for operators, including decreased profitability. This decline is largely due to the lack of attention to maximizing vehicle capacity for transport services, leading to inefficient use of resources [22]. Table 1 presents ways and means of solving problems related to the management of peak loads in public transport.

When solving problems related to ensuring traffic safety when transporting passengers, it is necessary to consider factors such as driver qualifications, structural safety, technical condition, and compliance with vehicle operating rules [9]. The negative consequences of accidents in the field of passenger services include human sacrifice, material damage, and decreased quality of life for the population. Table 2 provides extensive information on passenger safety and standards compliance, including solutions and real-life examples from different countries and cities.

Ensuring road safety is important when transporting passengers. Naturally, special attention must be paid to ensuring road safety. Let us consider the key factors affecting passenger transportation safety and how its organization depends on them. The first issue is ensuring compliance with standard safety requirements during the passenger transportation process.

The strategic solution to this issue is to organize passenger transportation processes in strict accordance with the established standards. The tactical solution will consist of regular inspections of vehicles for compliance with standards, the introduction of advanced security systems, and the provision of safe conditions for picking up and dropping off passengers. This requirement is provided differently in different countries. For example, in Japan, strict regulations governing public transport services and safety standards ensure a consistently high level of service quality and passenger safety.

The second problem is the training of employees and drivers. Professional training of drivers and employees is considered a strategic solution to this problem. The

Table 1 Ways to reduce peak loads and manage flows

No	Problem	Strategic solution	Tactical solutions	Examples
1	Analysis of behavioral patterns	Studying passenger behavior during peak hours, including information on departures and destinations	Data collection and analysis through big data technologies; applying machine learning algorithms to predict peak periods	Using machine learning algorithms in London to predict passenger flow on the Met
2	Flexible scheduling method	Arrangement of transport schedule. The organization of temporary routes and reduction of interval times helps reduce the "tension" of public transport	Integration of transport management systems with optimization algorithms; introduction of time-based variable tariffs to ease demand during peak hours	In Tokyo, a variable fare system reduces train congestion during peak periods and encourages passengers to travel during less busy hours
3	Integrated multimodal transport systems	Creation of coordination systems that provide passengers with flexible movement in the metro, buses, and trains	Development of mobile applications for planning multimodal routes; construction of transition nodes for the convenience of changing traffic	In Singapore, the multimodal transport card system makes it easy to switch between different modes of transport with a single ticket
4	Technological solutions	Use of advanced technologies, smart city systems, and mobile applications	Development of AI systems for flexible management of the movement schedule; introduction of electronic tickets and payment systems	In Seoul, a smart city system uses data from mobile apps to optimize bus schedules and reduce waiting times
5	Improvement of infrastructure strain	Development of infrastructure and construction of additional lanes for public transport	Creation of ecological corridors for cyclists and pedestrians; expansion of transport centers with an increase in the number of crossings	The construction of wide bicycle lanes in Beijing has reduced car traffic and made the city more environmentally friendly

(continued)

Table 1 (continued)

No	Problem	Strategic solution	Tactical solutions	Examples
6	Promotion of alternative modes of transport	Support for alternative modes of transport (e.g., cycling)	Introduction of benefit programs for cyclists; development of car-sharing system and bike-sharing network	Copenhagen's bike-sharing system and infrastructure for cyclists reduce traffic congestion and improve the city's ecology
7	Planning and forecasting the dynamics of population growth and mobility	Taking into account population growth and changes in population mobility when planning urban infrastructure	Analysis and forecasting of peak loads on public transport; application of geospatial modeling technologies	Shanghai uses geospatial data to optimize bus schedules to reduce congestion

Source Developed by the authors

tactical solution is to regularly check vehicles for compliance with standards, introduce advanced security systems, and ensure safe conditions for picking up and dropping off passengers. In Germany, public transport drivers are regularly trained and certified, including safety and driving psychology courses.

Technological security solutions appear as the third problem. A strategic solution to this issue lies in the use of advanced technologies in the field of security (e.g., video surveillance systems, driver monitoring systems, and emergency response systems). The tactical solution is to install video surveillance systems in vehicles, implement the driver's electronic control system, communicate with the dispatcher, and develop emergency communication programs. In this direction, the use of security cameras on buses used in London will increase the level of safety and help in investigating incidents.

The fourth problem lies in the emergency response systems. A strategic solution to this problem is rapid response to passenger emergencies. Emergency communication and braking systems are important in this case. The tactical solution to the problem lies in introducing emergency braking and warning systems for drivers, developing emergency communication systems with passengers, and training employees to act in emergencies. In Stockholm, emergency braking systems on subway trains allow for quick responses to potentially dangerous situations.

Passenger transport is part of the general traffic flow. Since passenger transport is dominant in the traffic flow, it is a source of environmental pollution. Along with other vehicles, passenger transport also emits harmful substances into the atmosphere (e.g., oxides of carbon, nitrogen, sulfur, and particulate matter). In turn, these substances cause smog, acid rain, and other environmental problems. The increasing number of vehicles in large metropolises is causing various problems. In Tashkent, traffic jams are a common issue on the highway networks. At some intersections, vehicles often back up 4–5 or more cycles away from the traffic lights, resulting in significant

Table 2 Ways to increase environmental safety of passenger transport

No	Problem	Strategic solution	Tactical solutions	Examples
1	Electric and hybrid vehicles	Use of vehicles powered by electricity or hybrid motors	Operating electric buses and providing them with charging stations or buses with hybrid engines	In the Chinese city of Shenzhen, which operates a large number of electric buses, it has led to a significant reduction in air pollution
2	Use of renewable energy sources	Integrate renewable energy into transportation systems to reduce dependence on oil resources and reduce greenhouse gas emissions	Introduction of alternative energy sources for public transport	In Stockholm, solar panels on the roofs of bus stops power traffic light systems
3	Application of new technologies to improve efficiency	Application of advanced technologies that provide savings in fuel consumption and emissions of harmful substances into the atmosphere	Implementation of waste recycling systems, development of technologies for more economical use of fuel, and the creation of lightweight vehicles through the use of polymer materials, which ensures their efficient operation	In Japan, internal combustion engines with the addition of a hydrogen mixture are used in buses to reduce the level of emissions of harmful substances into the atmosphere
4	Increasing the movement of bicycles by creating the necessary conditions for them and properly arranging pedestrian crossings	Creating the necessary conditions for the safe movement of cyclists and pedestrians	Increased construction of bicycle and pedestrian paths and sidewalks; expansion of bicycle rental programs; organization of pedestrian zones; taking the initiative to support such mobility	Copenhagen has created infrastructure and sufficient conditions to support the movement of bicycles, which helps reduce air pollution

Source Developed by the authors

delays. As a result, many citizens are suffering from the harmful gases emitted by vehicles while idling at these intersections [18].

The negative impact of public passenger transport on the environment is manifested in the following:

- With the wrong driving mode, public passenger transport emits more harmful substances into the atmosphere than with the correct choice of driving mode;

- Despite being more environmentally friendly than private vehicles, public passenger transport is not utilized as efficiently due to its lack of attractiveness;
- An outdated fleet of public passenger transport and the impossibility of quickly updating them sometimes force the operation of old rolling stock, which emits more harmful substances to the environment compared to new ones.

With the increasing impact of climate change and increasing environmental awareness, creating environmentally friendly passenger transport systems is becoming an important aspect of urban infrastructure development. Table 2 presents practical methods and examples of systems that increase the environmental safety of passenger transportation in large cities.

3 Results

New technologies make it possible to increase the safety, efficiency, and environmental safety of transport. In turn, the use of outdated systems and technologies and existing technological limitations in the organization of passenger transportation can lead to the following negative consequences:

- Safety of transport and high risk of technical malfunctions that can lead to accidents;
- Inefficiency of transport, low throughput of transport systems, and increased fuel consumption and emissions of harmful substances;
- Inconvenient use of transport services due to outdated payment and ticketing systems that can make it difficult for passengers to use transport services.

Based on the analyses and studies mentioned above, the authors developed the following proposals to address several issues and improve technological systems in passenger transportation:

1. Zero-emission electrification and transport. A strategic solution to this problem is the transition to electric energy transport and reducing the emissions of harmful substances. The tactical solution is to introduce electric buses and trains. It is necessary to develop networks of charging stations and apply hydrogen fuel cell technologies. As a solution to this problem, electric buses are in operation in Paris, while Germany has introduced trains powered by hydrogen fuel.
2. Mobile applications and digital platforms. A strategic solution to this issue is the development of mobile applications and digital platforms that allow passengers to plan routes and efficiently pay for services. It is recommended that route planning programs be developed and mobile payment systems introduced. Moreover, it is necessary to create digital platforms for feedback, which is a tactical solution to the problem. On this route, passengers in Tokyo use mobile apps to pay fares and keep up-to-date information.
3. Intelligent traffic management systems. A strategic solution to this problem involves using AI and smart traffic management systems, which optimize traffic

schedules, reduce congestion, and improve the efficiency of passenger transportation. As a tactical measure, the development of AI systems for traffic optimization is recommended, integrating sensors and cameras to monitor traffic flow. In London, intelligent traffic light control systems are employed to regulate and smooth traffic flows, effectively addressing congestion.

4. Autonomous vehicles. With the use of autonomous technologies in organizing passenger transportation, it is possible to create safer and more efficient systems, which helps reduce the human factor and, most importantly, reduce the risk of road accidents. This is a strategic solution to this problem. As a tactical solution, it is recommended to test and introduce autonomous buses and taxis and develop an uncrewed train system. Another proposal is to create infrastructure for autonomous transport. For example, Singapore is experimenting with autonomous taxis, which will allow passengers to use uncrewed vehicles.

Technological limitations make it necessary to continuously modernize outdated systems, requiring consistent investment. This enables transport service operators to adopt new technologies, improving the efficiency and safety of logistics systems. Additionally, for the strategic and systematic development of passenger transport in large urban areas, the following measures should be implemented:

- Strict compliance with the requirements of traffic safety standards and quality of service when organizing passenger transportation and their further improvement;
- Considering population growth and their well-being, which positively impact the transport mobility of the population. It is necessary to plan and forecast these indicators, which will ensure the use of logistics methods in organizing passenger transportation;
- Ensuring broad public participation in the development of projects for planning and development of logistics systems for passenger transportation. This makes it possible to consider the interests of different population groups and increase passenger satisfaction.

4 Conclusion

Thus, the research results indicate that the current state of passenger transportation organization and the degree of logistics method implementation are insufficient and require further improvement. This calls for the study of advanced practices and their systematic application within the industry. As the research demonstrates, organizing passenger transportation through logistics systems requires using various tools and technologies while considering multiple factors.

References

1. Abdullaev BI, Nazarov AA (2018) Ways quality increasing of exploitation of city bus capacity and traffic. Eur Sci Rev 5–6:374–376
2. Sh, Alimukhamedov, Nazarov A, Nazarov A, Ustoboev A (2022) Estimation of the interval of movement of public passenger transport in the direction. AIP Conf Proc 2432:030055. https://doi.org/10.1063/5.0090848
3. Andreev KP (2016) Safety of passenger road transport. New Sci Idea Result 12–3:14–17
4. Arifjanova NZ (2023) Modern trends in the development of urban passenger transport. Problems Modern Sci Educ 3(181):18–20
5. Arora A, Anand A, Banerjee-Ghosh S, Baraya D, Chakrabarty J, Chatterjee M (2016) Integrating intermediate public transport within transport regulation in a megacity: a Kolkata case study. New Delhi, India: Centre for Policy Research
6. Butaev ShA, Sidiqnazarov QM, Murodov AS, Ko'ziev AO (2012) Logistics (flow management in the supply chain). Tashkent, Uzbekistan: Extremum-Press LTD
7. Darmogray AV (2003) Logistics coordination process and system of urban passenger transport (Synopsis of Dissertation of Candidate of Economic Sciences). Yuri Gagarin State Technical University of Saratov, Saratov, Russia
8. Farahani RZ, Miandoabchi E, Szeto WY, Rashidi H (2013) A review of urban transportation network design problems. Eur J Oper Res 229(2):281–302. https://doi.org/10.1016/j.ejor.2013.01.001
9. Ganiev IG (2022) Problems of operation of road structures in the countries of the world. Sci Educ 3(10):214–225
10. Gudkov VA, Mirotin LB, Velmozhin AV, Shiryaev SA (2006) Passenger automobile transportation: Textbook. Moscow, Russia: Hotline-Telecom
11. Kandel BK, Martseniuk L, Marchenkova YK (2022) Passenger transport hubs: Thematic overview. Philos Econ Law Rev 2(1):76–87. https://doi.org/10.31733/2786-491X-2022-1-76-87
12. Kenzhaeva BO (2023) Factors affecting the efficiency of transport activity in cities. J New Century Innov 43(3):130–135
13. Malyshev MI (2022) Innovations in the field of urban public transport and prospects of implementing new mobility principles. Nauchnyi Vestnik MGTU GA [The Civil Aviation High Technologies] 25(3):36–50. https://doi.org/10.26467/2079-0619-2022-25-3-36-50
14. DYu, Melentyev (2011) City passenger transport: from total state monopolism to complete right of business structures. Econ Herald Donbas 2(24):12–16
15. Mironova AV (2015) Transport problem and its solution in the metropolis—an example of Moscow and New York: Comparative analysis. RUDN J Russian History 1:91–101
16. Mirzaev F, Nazarova V, Akhmedov D, Yusufkhonov Z (2022) The quality of logistics services and staff qualifications. Universum Techn Sci 2–3(95):28–31
17. Morton C, Caulfield B, Anable J (2016) Customer perceptions of quality of service in public transport: Evidence for bus transit in Scotland. Case Studies Transp Policy 4(3):199–207. https://doi.org/10.1016/j.cstp.2016.03.002
18. Nazarov AA, Usmonova MN, Rakhmat BA (2023) The role of legal culture of drivers in ensuring traffic safety. J Univ Sci Res 1(5):1190–1193
19. Nazarov A, Ustabaev A, Nazarov A (2022) Method of determination of public passenger transport interval for "critical situations. Harv Educ Scientific Rev 1(1):157–164. Retrieved from https://journals.company/index.php/hesr/article/view/19/19. Acessed 30 April 2024
20. Nazarov A, Ustabaev A, Nazarov A (2022) Selection of rational order of buses in traffic routes. Harvard Educ Scientific Rev 2(1):14–19. Retrieved from https://journals.company/index.php/hesr/article/view/21/21. Accessed 30 April 2024
21. Nazarov A, Ustaboyev A (2021) Criteria for assessing the provision of transport services to passengers in the urban community. Scientific J Veh Roads 2:23–26
22. Postnikov VP (2016) Analysis of the functioning of public passenger transport in large cities. Sci Technol Transp 2:100–107

23. Rustamov K, Rustamova N, Nazarov A, Abdukodyrova M (2022) The state of the transport services market and development trends. AIP Conf Proc 3045:060036. https://doi.org/10.1063/5.0197339
24. Shalaik B, Jacob R, Winstanley A (2010) Integrating real-time bus-tracking with pedestrian navigation in a journey planning system. In: Proceedings of GISRUK 2010: GIS Research UK 18th annual conference, pp 505–510. London, UK: University College London
25. Sorokin SV (2006) Increasing the efficiency of the functioning of urban passenger transport based on the redistribution of passenger flows (Synopsis of Dissertation of Candidate of Economic Sciences). Russia Siberian State Automobile and Highway University (SibADI), Omsk
26. Spirin IV (2006) Transportation of passengers by urban transport: Textbook. Moscow, Russia: Akademkniga
27. Zito P, Salvo G (2010) Toward an urban transport sustainability index: an European comparison. Eur Transp Res Rev 3(4):179–195. https://doi.org/10.1007/s12544-011-0059-0

Transport Services Trends in the Effective Use of Green Energy Sources in the Vehicle Network in the Context of Transformation

Komil B. Urazov, Alisher B. Fayzillayev, and Rayhona A. Artikova

Abstract The research aims to study the practical issues of revenue and cost management in the automotive industry and conduct a structural analysis of material costs. The authors use comparative analysis to identify the economic growth rates (according to the data of the motor transport company) in 2019–2022. The research object is a motor transport company. The effective use of green energy sources in the sustainable development of the vehicle network (i.e., the effective use of electric cars moving using electricity instead of gasoline, diesel, gas, and other fuels to move vehicles) has been ensured in every possible way. State regulation of the transport sector is implemented through various mechanisms, including licensing, certification, provision and notification procedures, technical regulation, the implementation of scientific and technical policies, taxation, and tariff policies—particularly for socially significant transport services. It also involves ensuring state oversight of compliance with transport legislation, managing personnel policies, and utilizing other regulatory methods as provided by law. Automotive companies create and use indicators that allow them to effectively manage income and expenses, group and register them, improve the quality of services, and increase efficiency. Based on the analysis of the dynamics of changes in accounting data for the income and expenses of a motor transport company, it is justified that the use of green energy will reduce current costs and increase the enterprise's profit.

Keywords Motor transport · Income · Expenses · Analysis · Accounting · Korhonalari motor transport · Balance sheet · Passenger transport · Freight transport

JEL codes B29 · D1 · L91 · O18 · R19 · R49 · J19 · M49

K. B. Urazov (✉)
Samarkand Institute of Economics and Service, Samarkand, Uzbekistan
e-mail: komil-ishonch@bk.ru

A. B. Fayzillayev · R. A. Artikova
Tashkent State Transport University, Tashkent, Uzbekistan

S. M. Sultanova et al. (eds.), *Sustainable Development of Transport*, Lecture Notes in Networks and Systems 1366, https://doi.org/10.1007/978-3-031-88846-5_8

1 Introduction

Transport is the basis of the state's economy. In terms of employment and the cost of fixed assets, it surpasses many sectors of the domestic economy. Transport stimulates economic globalization and, as a circulatory system, ensures the unity of the national economy. The Republic of Uzbekistan is taking targeted measures to develop its transport potential.

Since ancient times, humanity has relied on various means of transportation to move from one place to another. Initially, animals like horses, donkeys, and camels were used to carry passengers and goods, while oxen were employed to plow the land. As a result of technological development, automobiles were invented, gradually replacing horse and donkey carts for passenger and cargo transportation. This shift led to the rapid adoption of tractors, cranes, earthmoving machines, and other technologies in construction and earthworks. Currently, humanity has made unprecedented progress in motor transport, with increasingly efficient and advanced vehicles being produced each year. Given the rapid growth of the automobile industry, the study of vehicle classification, its descriptive aspects, and its treatment as accounting objects has become a highly relevant topic [4, 8, 9, 24].

2 Materials and Method

K. B. Urazov, B. A. Hasanov, N. B. Abdusalomova, and others published textbooks, study guides, and books about "Classificatory and descriptive bases of motor vehicles in objective quality." In his book "Peculiarities of accounting in other branches," K. B. Urazov expressed the following points based on the classification and description of motor vehicles as an object of accounting [19, 20], "transport" is a special branch of the national economy, which includes railway, air, river, sea, underground, and surface transport. Auto transport is a branch of the transport network specializing in transporting goods and passengers using cars. A vehicle transport enterprise is a legal entity specializing in transporting goods and passengers using vehicles. Work performed in motor transport enterprises (services provided) is the transportation of goods and passengers to residents and legal entities by means of motor vehicles (services provided).

Averkamp [5], Maltseva [17], Wilmsmeier [26], and others also published various books on the classificatory and descriptive basis of auto transport vehicles as an accounting entity. For example, H. Averkamp focused on a long-term asset account that reports a company's cost of automobiles, trucks, etc. [5]. The account is reported under the balance sheet classification property, plant, and equipment. Vehicles are depreciated over their useful lives [22].

3 Results

The stable and efficient operation of the transport sector is essential for achieving high rates of economic growth, safeguarding the integrity and national security of the country, enhancing the standard of living for the population, and facilitating the prudent integration of Uzbekistan into the global economy. Market relations and the competitive environment in Uzbekistan's road transport are poorly developed. The main systemic problems are the weak development of forms of market relations, mechanisms of transformation of ownership, and the competitive environment. The country has developed a draft "Strategy for the development of the transport system of the Republic of Uzbekistan until 2035," which provides conditions for the growth of volumes and quality of passenger and cargo transportation.

Motor vehicles play a leading role in the material and technical base of motor transport enterprises. Without them, motor transport enterprises cannot carry out their main activity, which consists of cargo transportation, passenger transportation, and other works (services) [16]. Therefore, it is important to have a sufficient number of motor vehicles in the motor transport enterprises, ensure their completeness, serviceability, and updating, and keep a strict account of the work activity of each vehicle.

Motor vehicles at the disposal of motor transport enterprises can be classified according to different characteristics as an accounting object.

According to the purpose of use, vehicles of motor transport enterprises are designed for passenger transportation, cargo transportation, and special work [13, 15, 23].

Vehicles intended for passenger transportation include buses, minibusses, and passenger cars used as taxis.

Examples of vehicles intended for cargo transportation are flatbed, van, tipper, tractor, and their trailers and semi-trailers.

Vehicles intended for special work (services) carry water, milk, fuel, bread, and other types of cargo. They can also be used for special work, for example, vehicles that carry out work such as lifting, breaking, and crushing, fire extinguishing, garbage transportation, cleaning asphalt roads from snow and ice, sweeping, sprinkling water, and others.

All types of these vehicles differ from each other according to the place of development, name, brand, and type of fuel consumed.

Nowadays, the use of electric cars is also growing year by year. Electric cars are vehicles that move using electricity and do not emit toxic gases into the atmosphere. The use of electric cars is becoming increasingly relevant. Particularly, electric cars worth $69.8 million were imported to the Republic of Uzbekistan in 2022 [21]. In the UK, sales of electric cars increased to a record level in July 2022 by 395,000 in 2022 (92%) from 2020 [6].

Motor vehicles are included in the balance sheet of motor transport enterprises according to their initial value.

According to the BHMS No. 5 "Main vehicles" of the Republic of Uzbekistan, the initial cost of motor vehicles includes the following:

- The purchase price of the vehicle;
- Costs of transportation of the vehicle;
- Customs fees on vehicles (regarding imported vehicles);
- Travel expenses paid to employees mobilized to deliver the vehicle to the enterprise;
- Fees for brokerage service;
- Expenses for the state registration of the vehicle (account certificate, technical passport, payment of 6% of the purchase price to the road fund);
- Other expenses related to the purchase and delivery of the vehicle.

The initial cost of vehicles must be calculated by the accounting department of the vehicle transport enterprise in the above manner. This calculation must be approved by the head of the enterprise.

To engage in transporting goods and passengers by vehicle, it is necessary to obtain a license. For this purpose, it is necessary to send an application through the "my.gov.uz" website to the section "License for the transportation of passengers and cargo on international routes." A license will be issued after the payment of the prescribed state duty.

Modernization of motor vehicles usually means work that leads to the improvement of their technical and operational characteristics (e.g., power, useful life, and other aspects), as well as changes in the field of application. Examples of modernization works include replacing an old engine in a motor vehicle with a more powerful one, lengthening or re-equipping its body, and adding additional springs. The costs associated with these upgrades are added to the initial cost of the motor vehicles [1, 4, 8].

Expenditures such as replacing vehicle tires and other spare parts or replacing an old engine with a new engine of the same capacity are not recognized as capital investments. Instead, these costs are written off as vehicle repair expenses [5, 14, 18].

Vehicle repair is divided into three types: current, medium, and major repairs.

Current maintenance is carried out continuously for preventive purposes. Current maintenance is usually carried out quarterly, as well as when malfunctions occur [10].

Intermediate repairs are carried out at least once a year. It differs from the current repair in that some parts and assemblies of motor vehicles are replaced [2, 7, 11, 12].

Capital repair work is carried out in different terms according to the business plans of motor transport enterprises. For example, motor vehicles may be repaired once in several years, depending on the technical condition and intensity of operation. The difference between capital repair and current and medium repair is that the motor, board, and other parts of the motor vehicle are completely removed and reinstalled, as well as completely replaced units that have become unusable [1, 3, 9, 23].

In the balance sheet of motor transport enterprises, motor vehicles are reflected in their residual value. The residual value of vehicles means the difference between their initial (restored) value and accumulated depreciation amounts (Table 1).

Table 1 Entering vehicles into the accounting balance sheet

The content of the operation	Amount	Debit	Credit	Basic document
1. To the value of fixed assets at the purchase price	100,000,000	0820	6010	Invoice, invoice, customs cargo declaration
2. To the amount of customs duties on fixed assets	50,000,000	0820	6990	Customs cargo declaration
3. To the amount of brokerage fee in the purchase	1,500,000	0820	6990	Invoice
4. To the amount of travel expenses on purchase	1,750,000	0820	6970	Prepayment bank receipt
5. To the amount of transportation costs	2,000,000	0820	6990	Invoice
6. To the amount of payments to the state traffic control	175,525	0820	6990	Invoice
7. 3% of the purchase price to the road fund	3,000,000	0820	6520	Payment order
8. Enter the vehicle at its initial cost	158,425,525	0160	0820	Final calculation

Source Developed and compiled by the authors

If we look at the statistics, transport services in the Republic of Uzbekistan and their share in the economy are increasing daily. As can be seen from Table 2, the volume of passenger transportation in the Republic of Uzbekistan increased from 1404.2 million people to 4541.3 million people in Q1, Q2, and Q3 2023 (Table 2 and Fig. 1).

The volume of cargo transportation has increased from 5124 billion soums to 41,727 billion soums in the last 12 years (i.e., from 2010 to 2022). In 2022, the gross domestic product of the Republic of Uzbekistan amounted to 896,617.9 billion soums. Thus, 4.65% of the GDP of the Republic of Uzbekistan fell on cargo transportation services. The importance of motor transport services in the Republic of Uzbekistan is high. However, the harmful gases coming out of these vehicles are polluting the atmosphere every day. For example, in the city of Tashkent on December 21, 2023, the amount of PM2.5 particles exceeded the permissible amount by 57 bars. Science has proven that PM2.5 particles cause cancer and asthma. According to UzHydroMet, on December 17 [25], the maximum content of small dispersed PM10 particles in the air of Tashkent was 241 $\mu g/m^3$, and the content of PM2.5 tiny particles was 211 $\mu g/m^3$.

These figures are 43 times more than WHO recommendations. The maximum content of carbon monoxide (gas) was 6.2 $\mu g/m^3$, which is 1.4 times more than the permissible level. These harmful gases are released into the atmosphere not only from motor vehicles. Harmful gases also come from industrial production plants. However, the main part of the harmful gases released into the atmosphere comes from motor vehicles.

Table 2 Volume of passenger transportation in the Republic of Uzbekistan in Q1, Q2, and Q3 2023

Region	Q1 2023	Q2 2023	Q3 2023	Q4 2023
The Republic of Uzbekistan	1404.2	2917.7	4541.3	6270.2
The Autonomous Republic of Karakalpakstan	32.9	68.4	106.3	146.8
Andijan Region	165.5	347.9	533.9	737.8
Bukhara Region	73.1	136.3	211.7	292.3
Gizzakh Region	27.4	43.7	67.9	93.7
Kashkadarya Region	69.6	164.2	241	333.5
Navoi Region	34.5	83.3	128.9	178.4
Namangan Region	107.7	212.9	331.3	457.5
Samarkand Region	107.5	203.8	316.5	436.9
Surandarya Region	57.5	135.2	210	290.0
Syrdarya Region	21.8	40.9	63.5	87.7
Tashkent Region	161.8	343.4	533.2	739.2
Ferghana Region	144.4	325.2	505.3	697.6
Khorezm Region	95.4	214.4	332.7	459.9
Tashkent city	305.1	598.2	959.1	1318.9

Source Compiled by the authors

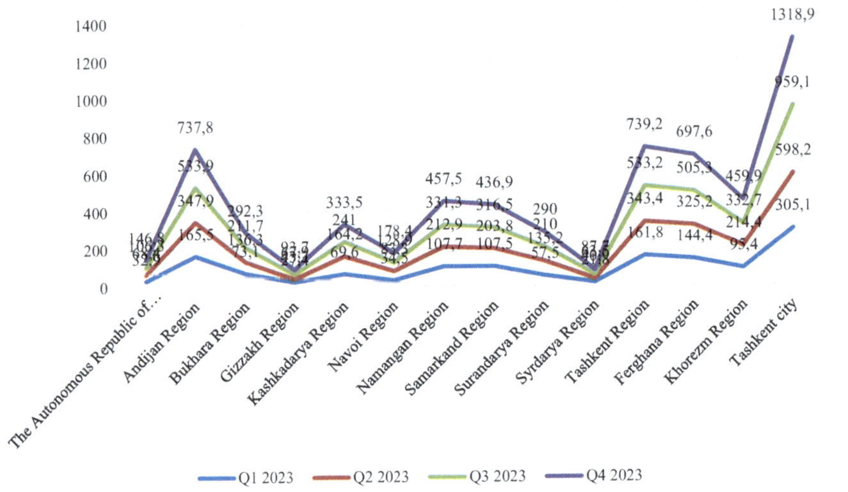

Fig. 1 The volume of passenger traffic in the Republic of Uzbekistan in Q1, Q2, Q3, and Q4 2023. *Source* Compiled by the authors

4 Conclusion

Motor vehicles play a vital role in socio-economic life. We are seeing good revenue from motor transport services. However, we should not spend the income we have seen to restore our health, which we are losing due to the particulate gases emitted by these vehicles, for treatment. For this purpose, it would be appropriate to do the following:

1. To increase the use of electric cars in cargo transportation and passenger transportation, prevent atmospheric pollution by producing electric cars ourselves, introduce paid parking lots for gas, gasoline, and diesel vehicles, and introduce free parking lots for electric cars;
2. To prevent the cutting of trees and set a large fine for cutting trees;
3. To effectively organize stockpiles of cargo and other material flows in the economy, it is necessary to increase the efficiency of the transport system and solve organizational and economic problems.

The authors emphasize the need to effectively organize the transport system's activities. This includes developing methodologies, conclusions, and recommendations to enhance the system's efficiency. Key aspects involve creating short-term and long-term strategies for the system's development, modernizing the infrastructure, ensuring effective coordination and integration with business partners, delivering high-quality service to customers, and boosting the competitiveness of logistics service providers. These measures collectively contribute to improving the overall efficiency of the transport system.

References

1. Abduganiyeva G, Yakhyaeva D (2019) Improve the recording and analyzing the income and expenses of enterprises. Bulletin Sci Pract 5(9):278–282. https://doi.org/10.33619/2414-2948/46/35
2. Abdurakhmonov RA (2019) Improvement of accounting of costs at enterprises. Bulletin Sci Pract 5(12):324–328. https://doi.org/10.33619/2414-2948/49/38
3. Abdusalomova NB (2022) Improving methodological approaches to the development of budget models in the management accounting system. Topical issues of transition to international financial reporting standards and application of international experience in Uzbekistan: Collection of materials international scientific and practical conference. Tashkent State University of Economics, Tashkent, Uzbekistan, pp 5–17
4. Akhmedov AM (2018) Development of cost management methods in the financial management system of a motor transport enterprise. In: Yakovleva TV (Ed), Scientific research: Theory, methodology, and practice, pp 308–314. Cheboksary, Russia: Interactive plus
5. Averkamp H (2016) Accounting coach. Retrieved from https://www.accountingcoach.com/. Accessed 15 February 2024
6. Averkamp H (2024) Vehicles definition. In: Accounting coach. Retrieved from https://www.accountingcoach.com/terms/V/vehicles. Accessed 13 August 2024

7. Bakieva I, Temirov F (2019) Prospects for the development of transport services in Uzbekistan and the theoretical basis for accounting for income and expenses in enterprises. Bulletin Sci Pract 5(12):329–335. https://doi.org/10.33619/2414-2948/49/39
8. Balas EA (1968) A note on the branch-and-bound principle. Oper Res 16(2):442–445
9. Dubovik II (2014) Vehicle operation. Transp Serv Account Taxation 6:33–35
10. Gavrilenko NG (2014) Innovative changes in road transport in the context of the cyclical development of the economic system. Vestnik SibADI 1(35):132–136
11. Gelyuta IV (2017) Peculiarities of formation of cost of transport services and their reflection in account of organization in transport sphere. Science. Thought 4:109–116
12. Juraeva G, Mirzaev N, Hamroev R, Rustamov M (2023) Transport features of logistics. E3S Web of Conf 402:01016. https://doi.org/10.1051/e3sconf/202340201016
13. Kalinina TB (2013) The methodology of cost accounting at motor transport enterprises. Issues Econ Law 60:66–69
14. Kalonov MB (2018) Regulation and incentive mechanism for foreign investors in Uzbekistan. In: Proceedings of the VI international scientific and practical conference social and economic aspects of education in modern society. Warsaw, Poland
15. Karakulov FZ (2019) Methods for analysis the problems of corporate governance of railway companies. Moscow Econ J 6:233–240. https://doi.org/10.24411/2413-046X-2019-16001
16. Kudelskaya AA, Belozertseva NP (2017) Features of the organization of the activities of a motor transport enterprise. In: Student science of the 21st century: materials of the XIII international student scientific-practical conference, pp 274–278. Cheboksary, Russia: Interactive Plus. https://doi.org/10.21661/r-116098
17. Maltseva TA (2016) Modern cost accounting and cost calculation systems (on the example of transport services). Topical issues of improving accounting, statistics and taxation of the organization: Materials of the V International scientific and practical conference. Tambov State University named after G.R. Derzhavin, Tambov, Russia, pp 111–119
18. Mammadov HS, Chebotareva ZV (2016) The influence of the specificity of motortansport branch on income and expense calculation. Vestnik universiteta [University Bulletin] 5:138–143. Retrieved from https://cyberleninka.ru/article/n/vliyanie-spetsifiki-avtotransportnoy-otrasli-na-uchet-dohodov-i-rashodov. Accessed 12 March 2024
19. Nobes C (2020) A half-century of accounting and business research: the impact on the study of international financial reporting. Account Bus Res 50(7):693–701. https://doi.org/10.1080/00014788.2020.1742446
20. Ravshanov MN (2012) Development of the road transport complex of the Republic of Uzbekistan. Econ Revival Russia 1(31)
21. Spicer & Co. (2022) 5 accounting issues for electric vehicles. Retrieved from https://www.spicerandco.co.uk/5-accounting-issues-for-electric-vehicles/. Accessed 13 August 2024
22. Statistics Agency under the President of the Republic of Uzbekistan (2024) Electric car. Retrieved from https://stat.uz/uz/component/search/?searchword=elektromobil&searchphrase=all&Itemid=114. Accessed 13 August 2024
23. Urazov KB, Sattarov TT (2022) Methodological issues of giving an audit assessment of the correct reflection of the net assets of enterprises in financial statements. In: "Problems of taxation and accounting in the context of the transition to the international financial reporting standards" Republican scientific practical conference on the topic, pp 510–520. Tashkent, Uzbekistan
24. Urazov KB, Vakhidov SV (2011) Features of accounting in other industries: textbook. Tashkent, Uzbekistan: Adib
25. UzHydroMet (2024) O'zbekiston respublikasida kutilayotgan ob-havo ma'lumotlari 30.10.2024. Retrieved from https://hydromet.uz/. Accessed 30 October 2024
26. Wilmsmeier G, Roser G, Alderman M, Ilina E (2014) Efficiency—a key ingredient towards sustainable supply chains. UNELCAC Bulletin FAL 331(3):1–10

A Systemic Approach to Managing Sustainable Development in the Fuel and Energy and Transport and Logistics Sectors Within the Regional Economy of Russia: An ESG Perspective

Ekaterina A. Zemlyacheva, Agnessa O. Inshakova, and Denis E. Matytsin

Abstract The research focuses on defining the ESG perspective in implementing a systemic approach to managing the sustainable development of the fuel and energy (FES) and transport and logistics sectors (TLS) in regions, using the regional economy of Russia as an example. The research is based on empirical data from the top 15 regions of Russia, which were leaders in the national ESG ranking in 2023. Through correlation analysis, the authors justify the preference for systemic management of sustainable development in the fuel and energy and transport and logistics sectors in Russian regions because this approach generates a more pronounced effect, reflected in the improvement of regions' positions in the ESG ranking. The authors conducted the regression analysis to develop an econometric model that revealed the causal relationships between the implementation of management measures in these sectors and the sustainable development of a region's economic system. This model identified a Pareto-optimal combination of selected management measures, ensuring the full sustainability of Russian regions by 2030. The main conclusion of this research is that the sustainable development of regions in Russia is achieved through the comprehensive implementation of management measures aimed at increasing the turnover of organizations in the fuel and energy sector, boosting turnover in the transport and logistics sector, and attracting investments into fixed capital in both sectors. This led to the formulation of scientific principles for a systemic approach to managing the sustainable development of these sectors in Russia's regional economy, with theoretical significance in advancing management tools for implementing the

E. A. Zemlyacheva
RUDN University, Moscow, Russia

A. O. Inshakova (✉) · D. E. Matytsin
Volzhsky Institute of Economics, Pedagogy and Law, Volzhsky, Russia
e-mail: ainshakova@list.ru

Kuban State University, Krasnodar, Russia

D. E. Matytsin
e-mail: dmatytsin@yandex.ru

S. M. Sultanova et al. (eds.), *Sustainable Development of Transport*, Lecture Notes in Networks and Systems 1366, https://doi.org/10.1007/978-3-031-88846-5_9

ESG agenda in regions. The proposed recommendations for realizing the ESG perspective on the sustainable development of regional economic systems in Russia by 2030, within the framework of Pareto optimization, can enhance the management practices of regional fuel and energy and transport and logistics sectors in Russia.

Keywords Systemic approach · Regional management · Sustainable development · Fuel and energy sector (FES) · Transport and logistics sector (TLS) · Regional economy of Russia · ESG perspective

JEL codes Q01 · Q43 · R41 · R58

1 Introduction

The fuel and energy sector (FES) and the transport and logistics sector (TLS) play a vital role in providing infrastructure for the region's economy. These sectors are closely interconnected. First, the ability to transport goods and ensure the smooth operation of passenger transport directly depends on the availability of the necessary fuel and energy resources. Environmentally friendly regional public and private electric transport systems, such as electric vehicles, rely on the uninterrupted supply of affordable electricity. The transition to autonomous vehicles further increases the energy demands of the TLS and intensifies the load on the FES. The availability and cost of automotive fuel determine the ability to deliver consumer goods to the region's logistics centers and local retail outlets.

Second, fuel and energy resources require safe and highly efficient transport logistics for their careful storage and successful transportation, which ensures profitable sales and energy security for the region's economy. Energy loss during distribution poses the risk of shortages and increases the final price. The leakage of fossil fuels (natural gas and oil) during transportation brings the risk of environmental disasters and public safety concerns because these fuels are highly explosive. To address this, regions are developing smart power grids (Smart Grids) and adopting advanced practices for the safe transport logistics of fuel and energy resources.

Third, economic activities in both sectors of the regional economy generate significant production and consumption waste. The transition to clean energy in a region makes it impossible to maintain transportation reliant on internal combustion engines. Similarly, the region's shift to electric transport is unfeasible in the face of electricity shortages or high costs. This highlights the necessity for joint implementation of green investment and innovation projects in the FES and TLS, as undertaking such projects independently in these sectors is inefficient or impossible.

The issue lies in the fact that the existing approach to managing the sustainable development of regional economies involves regulating the FES and TLS separately. The imbalance in the development of these sectors makes regions more vulnerable to economic, social, and environmental crises. ESG practices for managing sustainable development offer broad opportunities for integrated crisis management of economic systems. This underscores the importance of developing management measures to address the imbalance in the development of the FES and TLS in regions based on ESG principles. This rationale shaped the goal of this research, which is to define the ESG perspective for implementing a systemic approach to managing the sustainable development of the FES and TLS in regions, using the regional economy of Russia as an example.

2 Literature Review

The fundamental basis for the research is the theory of sustainable development in regional economies [7, 10, 11]. According to this theory, the sustainable development of a region's FES involves the continuous production of fuel and energy resources and their highly efficient distribution within the socio-economic environment of the region under conditions of universal accessibility, high utilization of production capacities in regional FES enterprises, and the overall financial and economic efficiency of these enterprises [1].

In line with this theory, the sustainable development of the TLS in a region implies the availability of various modes of transport and the ability to flexibly combine them when establishing logistics routes, the extensive reach of the regional transport logistics network, the safety and speed of transport services, high logistical efficiency, and the low cost of transport and logistics services [15, 16].

The primary reflection of the sustainability of a region's economic system lies in its adherence to the ESG management concept, which harmonizes society, economy, and the environment, while simultaneously addressing environmental (E), social (S), and governance (G) crises through the implementation of public and private management measures [4, 9, 14].

The following are key forward-looking management measures for sustainable development in a region's FES and TLS:

- Increasing the number of organizations to promote competition, expand regional production capacities, and create additional jobs in the FES [8] and TLS [3] within the region;
- Enhancing the turnover of organizations to improve the financial and economic performance and increase the utilization of production capacities in the region's FES [6] and TLS [17];
- Attracting investment in fixed capital to enable digital modernization and the introduction of advanced technologies in the FES [12] and TLS [2] of smart regions.

Despite the relatively thorough development of the listed management measures in existing publications, they do not explain the impact of implementing these measures on the sustainability of the region's economic system, specifically the extent to which the region's development aligns with the ESG management concept. This represents a gap in the literature and raises the research question of how to manage the FES and TLS in Russian regions to ensure their sustainable development and support the practical implementation of the ESG agenda in regional economies.

3 Materials and Methods

This research is based on the best practices of the top 15 regions of Russia, which were leaders in the national ESG ranking at the end of 2022, as of early 2023. The differences in the isolated and combined contributions of FES and TLS management measures to the sustainable development of regional economic systems are determined through correlation analysis, with correlation coefficients calculated and compared. The impact of FES and TLS management measures in the selected Russian regions (based on statistics of the data from the Federal State Statistics Service of the Russian Federation [Rosstat] for the end of 2022 [5], relevant as of early 2023) on their ESG ranking is determined using regression analysis. The data for the research are presented in Table 1.

Based on the results of the regression analysis, the authors selected the management measures in the FES and TLS that support regional sustainability, demonstrating positive regression coefficients in the ESG model. A Pareto-optimal combination of the selected management measures is identified, ensuring full sustainability (ESG = 1) of the regional economic systems of Russia by 2030 (at the end of the Decade of Action).

4 Results

The results of the correlation analysis of the data from Table 1 showed that when management measures are implemented separately in each of the studied sectors, the correlation of sustainable development in the regional economic systems of Russia in 2023 is 70.83% with the FES and 71.92% with the TLS. When management measures are implemented jointly in the FES and TLS, the correlation with the sustainable development of regional economic systems is significantly higher, reaching 82.53%. Therefore, it is advisable to apply a systemic approach to managing the sustainable development of the FES and TLS in Russian regions. To determine the impact of management measures in the FES and TLS in the selected Russian regions on their ESG ranking in 2023, the authors conducted a regression analysis of the data from Table 1, with the results presented in Table 2.

In Table 2, the significance of F is 0.0864. Thus, the results obtained from the regression analysis are reliable at a significance level of 0.10. This indicates the reliability of the regression analysis results and makes it possible to formulate the following econometric model for the dependence of the ESG ranking of Russian regions in 2023 on the management of sustainable development in their FES and TLS:

Table 1 Statistics of FEC, TLC, and ESG in Russian regions in 2023

Regions	ESG ranking, shares from 1	Number of organizations, thousand pcs		Turnover of organizations, billion rubles		Investments in fixed capital, trillion rubles	
		FES	TLC	FES	TLC	FES	TLC
	ESG	A_{FEC}	A_{TLC}	T_{FEC}	T_{TLC}	I_{FEC}	I_{TLC}
Moscow	0.781	1.3100	27.7240	3.9901	6.3060	0.1418	0.8043
Belgorod region	0.702	0.0880	1.8250	0.0753	0.0172	0.0054	0.0076
St. Petersburg	0.683	0.6560	13.4710	0.6829	1.3805	0.0645	0.1715
Kaliningrad region	0.630	0.1290	2.9240	0.0381	0.0463	0.0059	0.0200
Krasnodar Territory	0.662	0.6080	6.3510	0.3244	0.6322	0.0602	0.1389
Republic of Tatarstan	0.769	0.3860	5.8050	0.3570	0.3137	0.0495	0.1439
Tyumen region	0.723	0.5150	6.1280	0.4456	0.9259	0.0324	0.3704
Yamal-Nenets autonomous Area	0.674	0.0960	0.9350	0.0360	0.0733	0.0078	0.2908
Khanty-Mansi autonomous Area—Yugra	0.630	0.2250	2.9750	0.3616	0.6680	0.0155	0.0451
Sverdlovsk region	0.627	0.6320	7.5700	0.2166	0.3289	0.0318	0.0830
Novosibirsk region	0.619	0.5120	6.3580	0.1299	0.4805	0.0089	0.0403
Krasnoyarsk territory	0.605	0.4200	4.1260	0.3015	0.2758	0.0386	0.0776
Sakhalin region	0.649	0.0980	0.9750	0.0226	0.0786	0.0054	0.0486
Magadan region	0.647	0.0450	0.3480	0.0211	0.0199	0.0105	0.0057
Khabarovsk territory	0.645	0.2520	2.9730	0.1120	0.1434	0.0101	0.0738

Source Compiled by the authors based on the National Rating Agency [13] and Rosstat [5]

$$\begin{aligned} ESG =\ & 0.6424 - 0.1912 * A_{FEC} + 0.0105 * A_{TLC} + 0.0607 * T_{FEC} \\ & - 0.0773 * T_{TLC} + 0.9874 * I_{FEC} + 0.2547 * I_{TLC} \end{aligned} \quad (1)$$

Model (1) shows that an increase in the number of organizations in the FES leads to a decrease in the ESG ranking of Russian regions in 2023 by 0.1912. An increase in the number of organizations in the TLS contributes to an increase in the ESG ranking of Russian regions in 2023 by 0.0105. An increase in the turnover of organizations

Table 2 Regression statistics characterizing the impact of FES and TLS management measures in Russian regions on their ESG ranking in 2023

Regression statistics						
Multiple R	0.8253					
R^2	0.6812					
Normalized R^2	0.4420					
Standard error	0.0398					
Observations	15					
Variance analysis						
	df	*SS*	*MS*	*F*	*Significance F*	
Regression	6	0.0270	0.0045	2.8486	0.0864	
Residual	8	0.0126	0.0016			
Total	14	0.0397				
	Coefficients	*Standard error*	*t-statistics*	*P-value*	*Lower 95%*	*Upper 95%*
Y-intercept	0.6424	0.0214	30.0472	$1.6*10^{-9}$	0.5931	0.6917
A_{FEC}	−0.1912	0.1203	−1.5897	0.1506	−0.4686	0.0862
A_{TLC}	0.0105	0.0094	1.1224	0.2942	−0.0111	0.0321
T_{FEC}	0.0607	0.1299	0.4676	0.6525	−0.2388	0.3602
T_{TLC}	−0.0773	0.0907	−0.8515	0.4192	−0.2865	0.1320
I_{FEC}	0.9874	1.0467	0.9434	0.3731	−1.4263	3.4011
I_{TLC}	0.2547	0.1231	2.0701	0.0722	−0.0290	0.5385

Source Developed by the authors

in the FES results in an increase in the ESG ranking of Russian regions in 2023 by 0.0607.

An increase in the turnover of organizations in the TLS caused a decrease in the ESG ranking of Russian regions in 2023 by 0.0773. The inflow of investments into fixed capital in the FES led to an increase in the ESG ranking of Russian regions in 2023 by 0.9874. An increase in the volume of investments into fixed capital in the TLS leads to a rise in the ESG ranking of Russian regions in 2023 by 0.2547. Positive regression coefficients are found for the factor variables A_{TLC}, T_{FEC}, I_{FEC}, and I_{TLC}.

Based on the model, the following scientific principles of a systemic approach to managing the sustainable development of the FES and TLS in the regional economy of Russia are formulated:

- Encouraging the growth of competition and the creation of additional jobs by TLS organizations in the region, considering the current needs of the regional FES;

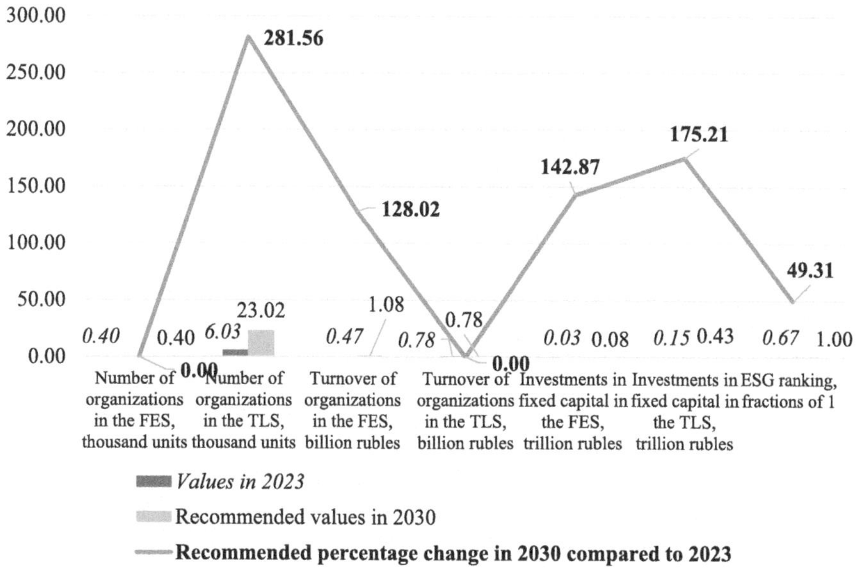

Fig. 1 The Pareto-optimal combination of selected management measures in the FES and TLS, ensuring full sustainability of the Russian regions by 2030. *Source* Developed by the authors

- Providing subsidies and tax support to improve the financial and economic performance of FES organizations in the region, focusing on the changing needs of the regional TLS;
- Organizing joint digital modernization and the implementation of advanced technologies in the FES and TLS of smart regions.

Based on the model (1), a Pareto-optimal combination of selected management measures in the FES and TLS has been identified, ensuring full sustainability (ESG = 1) of the economic systems of Russian regions by 2030 (at the end of the Decade of Action) (Fig. 1).

Figure 1 reveals the ESG perspective for the sustainable development of the regional economic systems of Russia by 2030. To practically implement this perspective, which aims to improve the ESG ranking of Russian regions by 49.31%, the authors propose a comprehensive set of management measures for regional FES and TLS, including the following:

- Increasing the turnover of organizations in the FES by 281.56%, from 6.03 thousand units in 2023 to 23.02 thousand units by 2030;
- Increasing the turnover of organizations in the TLS by 128.02%, from 0.47 billion rubles in 2023 to 1.08 billion rubles by 2030;
- Attracting investments in fixed capital in the FES by 142.87%, from 0.03 trillion rubles in 2023 to 0.08 trillion rubles by 2030;
- Increasing the volume of investments in fixed capital in the TLS by 175.21%, from 0.15 trillion rubles in 2023 to 0.43 trillion rubles by 2030.

5 Discussion

The contribution of this research to the literature lies in advancing the theory of sustainable development in regional economies [7, 10, 11] by providing a more comprehensive explanation and quantitative measurement of the effects of implementing FES and TLS management measures on the sustainability of regional economic systems in modern Russia, specifically the degree to which regional development aligns with the ESG management concept.

The authors found that an increase in the number of organizations in the TLS (as confirmed by Bogoviz et al. [3]), higher turnover in FES organizations (as confirmed by Gulnora [6]), an inflow of investments into fixed capital in the FES (as confirmed by Morozova et al. [12]), and an increase in fixed capital investments in the TLS (as confirmed by Audonin et al. [2]) ensure the sustainable development of regions in Russia.

However, it was also established that an increase in the number of organizations in the FES (unlike the findings of Karlibaeva and Nyangarika [8]) and growth in the turnover of organizations in the TLS (contrary to Tursunov [17]) do not support, as previously assumed, but rather hinder the sustainable development of regional economic systems in Russia. The results complement existing studies by Ergasheva et al. [4], Kitsai et al. [9], and Rogachev et al. [14] on the ESG agenda in regional economies, continue the series of studies by Alekseev et al. [1] on managing sustainable development in the FES in regions and contribute to the scientific discussion by Sultanova and Babakhanova [15, 16] on managing sustainable development in the TLS within regional economies.

6 Conclusion

Thus, the key results of the presented research, based on the progressive experience of the top 15 regions of Russia that led the national ESG ranking in 2023, are as follows:

(1) The preference for a systemic approach to managing the sustainable development of the FES and TLS in Russian regions has been substantiated, as in this case, management measures generate a more pronounced effect, reflected in the improvement of regions' positions in the ESG ranking;
(2) An econometric model was developed that revealed the causal relationships between the implementation of management measures in the FES and TLS and the sustainable development of the region's economic system;
(3) A Pareto-optimal combination of selected management measures in the FES and TLS was identified, ensuring full sustainability of Russian regions by 2030.

The overall conclusion is that the sustainable development of regions in Russia is achieved through the comprehensive implementation of management measures aimed at increasing the turnover of organizations in the FES, boosting the turnover of organizations in the TLS, and attracting investments in fixed capital in both sectors. The formulated scientific principles of a systematic approach to managing the sustainable development of the FES and TLS in the regional economy of Russia are theoretically significant, as they advance management tools for implementing the ESG agenda in modern regions.

The identified perspective on the sustainable development of regional economic systems in Russia by 2030 has practical significance, as it opens up opportunities for precise planning and forecasting of the implementation of the ESG agenda in Russian regions during the Decade of Action. The proposed recommendations for realizing the ESG perspective on the sustainable development of regional economic systems in Russia by 2030, within the framework of the Pareto optimization, are of managerial importance because they make it possible to improve management practices for regional FES and TLS in Russia.

References

1. Alekseev AN, Lobova SV, Bogoviz AV, Ragulina YV (2019) Digitalization of the Russian energy sector: state-of-the-art and potential for future research. Int J Energy Econ Policy 9(5):274–280. https://doi.org/10.32479/ijeep.7673
2. Audonin A, Turginbayeva A, Askerov A, Yergobek D (2020) Modern economic and logistical trends in Eurasia: how do new trans-Eurasian mega-projects influent to national economic growth. E3S Web Conf 159:06009. https://doi.org/10.1051/e3sconf/202015906009
3. Bogoviz AV, Ragulina YV, Lobova SV, Zhukov BM, Stepanova OM (2016) Services infrastructure forming in the process of transport logistics stock movement. Int Rev Manag Mark 6(6):278–283
4. Ergasheva ST, Zinovyeva IS, Abdurashitov AA, Kopytina YA, Makarova TV (2023) ESG investments in support of the development of the green economy in Russia and Central Asia. In: Popkova EG, Sergi BS (eds) ESG management of the development of the green economy in Central Asia. Springer, Cham, Switzerland, pp 419–428. https://doi.org/10.1007/978-3-031-46525-3_44
5. Federal State Statistics Service of the Russian Federation (Rosstat) (2024) Regions of Russia: socio-economic indicators 2023. Rosstat, Moscow, Russia. https://rosstat.gov.ru/storage/mediabank/Reg_Rus_Pokaz_2023.htm. Accessed 22 Aug 2024
6. Gulnora A (2022) Financial development and renewable energy consumption. In: Salahodjaev R (ed) Economic growth and wellbeing: evidence from the belt and road initiative countries. Nova Science Publishers, Hauppauge, NY, pp 35–53
7. Inshakova AO, Sozinova AA, Litvinova TN (2021) Corporate fight against the covid-19 risks based on technologies of industry 4.0 as a new direction of social responsibility. Risks 9(12):212. https://doi.org/10.3390/risks9120212
8. Karlibaeva RH, Nyangarika A (2023) Overview of investments in the energy and aluminum markets. In: Dinçer H, Yüksel S (eds) Renewable energy investments for sustainable business projects. Emerald Publishing Limited, Leeds, UK, pp 153–164. https://doi.org/10.1108/978-1-80382-883-120231012
9. Kitsai JA, Karpushova SE, Petrenko ES, Smetanina AI (2023) Social entrepreneurship as an institute of sustainable development risk management. In: Popkova EG (ed) Sustainable development risks and risk management. Springer, Cham, Switzerland, pp 363–367. https://doi.org/10.1007/978-3-031-34256-1_63

10. Matytsin DE, Inshakova AO, Goncharov AI (2023) Transformation of settlement relations in the context of industry 4.0: Conversion of blockchain club's crypto-codes into legitimate analogues. In: Popkova EG (ed) Sustainable development risks and risk management. Springer, Cham, Switzerland, pp 43–46. https://doi.org/10.1007/978-3-031-34256-1_8
11. Mkrtchyan TM, Tovmasyan GR, Dallakyan SA (2023) Sustainable development of the tourism sector of the Republic of Armenia in the context of an innovative economy. In: Popkova EG (ed) Smart green innovations in industry 4.0, pp 383–391. https://doi.org/10.1007/978-3-031-45830-9_43
12. Morozova IM, Litvinova TN, Przhedetskaya NV, Sheveleva VV (2018) The problems of financing of entrepreneurship infrastructure in developing countries and their solutions. In: Popkova E (ed) The impact of information on modern humans. Springer, Cham, Switzerland, pp 277–283. https://doi.org/10.1007/978-3-319-75383-6_35
13. National Rating Agency (2022) Ranking of sustainability of development of the Russian Federation and integration of ESG-criteria in the activities of the subjects of the Russian Federation: Analytical review. https://www.ra-national.ru/wp-content/uploads/2022/12/ranking_esg_regions.pdf. Accessed 22 Aug 2024
14. Rogachev AF, Ostrovskaya VN, Natsubidze AS, Litvinova TN, Yakovleva EA (2018) Tools for sustainability management of socio-ecological systems in the globalizing world. In: Popkova E (ed) The impact of information on modern humans. Springer, Cham, Switzerland, pp 241–247. https://doi.org/10.1007/978-3-319-75383-6_31
15. Sultanova SM, Babakhanova NU (2023) Statistical model of operational costs indicators in the intelligent transport system of JSC "Uzbekistan railways." AIP Conf Proc 2612:060014. https://doi.org/10.1063/5.0119125
16. Sultanova SM, Babakhanova NU (2023b) Statistical model for determining the quality of cargo work JSC "Uzbekistan railways. E3S Web Conf 402:06016. https://doi.org/10.1051/e3sconf/202340206016
17. Tursunov B (2022) Financial security in small business in period of digital economy: in case of Uzbekistan. In: Proceedings of ICFNDS '22: the 6th international conference on future networks & distributed systems. Association for Computing Machinery, New York, NY, pp 491–498. https://doi.org/10.1145/3584202.3584275

Digital Technologies and Infrastructure Support for the Sustainable Development of Transport and Logistics Systems

Modeling Traffic Flow Using the LWR Model and Finite Difference Numerical Method: Analysis of Efficiency and Applicability in Urban Transport Infrastructure Conditions

Makhamadaziz M. Rasulmukhamedov, Komoliddin Sh. Tashmetov, and Timur Sh. Tashmetov

Abstract The paper presents a research analysis of finite difference methods for the numerical solution of equations describing traffic flow on the example of the Lighthill–Whitham–Richards (LWR) model. The authors presented the main stages of finite difference methods, including the discretization of spatial and temporal variables and various types of finite differences. The authors analyzed different types of problems (the initial value problem and the boundary value problem) and methods for their solution. Additionally, the authors described how data on road traffic on the Tashkent Ring Road was collected using video detectors and general traffic characteristics. The results of numerical modeling of traffic flow based on the collected data and analysis of its effectiveness are presented. As a result, the authors confirmed the effectiveness of using finite difference methods for modeling traffic flow. The results also indicate the importance of vehicle speed in changing traffic density and show that density decreases rapidly as speed increases.

Keywords Traffic flow modeling · LWR model · Microscopic models · Finite difference method · Traffic density · Traffic speed · Numerical solution

JEL codes C6 · C8

1 Introduction

Mathematical models of traffic flow play a crucial role in effective transportation infrastructure management. They make it possible to analyze and predict vehicle flows, optimize road networks, and facilitate efficient utilization of resources. Modeling traffic flow aids in enhancing road safety, reducing congestion, optimizing

M. M. Rasulmukhamedov · K. Sh. Tashmetov (✉) · T. Sh. Tashmetov
Tashkent State Transport University, Tashkent, Uzbekistan
e-mail: tashmetov1993@gmail.com

S. M. Sultanova et al. (eds.), *Sustainable Development of Transport*, Lecture Notes in Networks and Systems 1366, https://doi.org/10.1007/978-3-031-88846-5_10

routes, and developing effective strategies for managing transportation infrastructure. Such mathematical approaches help in making informed decision-making to improve mobility and save time and resources, which is important for urban and intercity transportation systems [6].

There are three main types of mathematical models in traffic flow, each with unique characteristics: macroscopic, microscopic, and mesoscopic models.

Macroscopic models of traffic flow analyze road networks at the level of generalized characteristics, such as traffic density, total flow, and average speed. These models are designed for high-level understanding and optimization of road systems as a whole. They make it possible to investigate overall transportation infrastructure performance, forecast potential congestion, and optimize infrastructure planning [2, 6, 16].

On the contrary, microscopic models provide a detailed description of the behavior of particular vehicles. These models consider various aspects such as driving style, route preferences, and interaction with other vehicles. Microscopic models make it possible to accurately simulate complex scenarios (e.g., overtaking, wave effects, and interaction with traffic signals). They are suitable for studying the dynamics of traffic flow at the level of specific road sections or even individual intersections [3, 6, 16].

A mesoscopic model in the field of transportation modeling represents an intermediate level of detail between macroscopic and microscopic models. This model focuses on the level of groups of vehicles or road sections, providing more details on their interactions than the macroscopic models provide, but without the high level of detail typical of microscopic models [1, 6, 8, 12, 16].

The choice between traffic flow models depends on the research objectives. Macroscopic models are suitable for general system analysis at the city level. Mesoscopic models can be used for detailed examination of limited road sections. Microscopic models can be used when considering individual vehicle behavior at the level of specific road segments or intersections. The level of detail and the scale of the study are key factors in choosing the appropriate model for a particular task [6, 16, 17].

The authors conducted this research using a macroscopic traffic flow model. This approach allows for the examination of aggregate movement characteristics (e.g., density, flow, and speed at the system level). The analysis of the macroscopic model provides a general understanding of the effectiveness and capacity of road infrastructure, which can be valuable in optimizing urban or regional transportation systems.

2 Materials and Method

The macroscopic model was first introduced by Lighthill and Whitham in 1955 and independently by Richards in 1956. The mathematical equation they proposed, known as the Lighthill–Whitham–Richards (LWR) model, is a mathematical model of traffic flow that describes macroscopic characteristics of vehicle movement on a road section. The main idea is that changes in the density of vehicles on the road affect their speed and, consequently, the flow [6, 14, 16].

The LWR model includes the equation of mass conservation (1), which describes how changes in vehicle density affect the flow. It also includes the relationship between traffic speed and density (2). This relationship is typically expressed as a nonlinear function, reflecting the inverse relationship between density and speed [1, 3, 13, 15].

$$\frac{\partial p}{\partial t} + \frac{\partial (p \cdot v(p))}{\partial x} = 0 \tag{1}$$

$$v(x, t) = v(p(x, t)) \tag{2}$$

The primary numerical characteristics of traffic flow were first proposed in Bruce Greenshields' work. These characteristics include the average speed of movement (denoted as v) in kilometers per hour, the density of vehicles (denoted as p) in vehicles per kilometer, and the flow of vehicles (denoted as Q) in vehicles per hour. A relationship between these characteristics (3) was also proposed. This relationship is analogous to the relationship between similar quantities in fluid dynamics [4, 5, 7, 9].

$$Q = p \cdot v(p) \tag{3}$$

Thus, this work proposed important numerical characteristics for describing traffic flow and established the fundamental relationship between flow, density, and average speed, which is analogous to the principle of mass conservation in fluid dynamics. These concepts are fundamental to traffic flow theory. They make it possible to deeply understand its characteristics and interrelationships.

Additionally, Bruce Greenshields proposed the following formula for traffic flow velocity:

$$v = v_{max} \cdot \left(1 - \frac{p}{p_{max}}\right) \tag{4}$$

where v is the velocity, v_{max} is the maximum velocity, p is the density of vehicles, and p_{max} is the maximum density [1, 7, 9, 10].

This formula expresses a nonlinear relationship between the velocity of traffic flow and the density of vehicles. When there are no vehicles ($p = 0$), the velocity reaches its maximum (v_{max}). As the density (p) increases, the velocity decreases, and when the density reaches its maximum value (p_{max}), the velocity becomes zero [11].

Finite Difference Methods (FDM) are numerical methods used to solve equations and systems of equations by replacing derivatives with difference relations. These methods are widely used in the field of computational mathematics to approximate differential equations and solve various problems [6].

The main stages of solving partial differential equations with the finite difference method are as follows:

- Discretization of space and time.
- Approximation of derivatives.
- Statement of difference equations.
- Solving a system of equations.

Discretization is an important stage in finite difference methods, during which continuous spatial and temporal variables are replaced by discrete values on a grid. This process makes it possible to transform differential equations into difference equations and facilitates numerical solutions.

Spatial and temporal grids are typically represented as follows:

- $x_i = i \cdot \Delta x$—spatial grid, where i is the index of the grid node and Δx is the spatial step.
- $t_j = j \cdot \Delta t$—temporal grid, where j is the index of the grid node and Δt is the time step.

There are several types of finite differences depending on which nodal points are used for their computation. Right, left, and central finite differences are distinguished.

Finite-difference expressions for a particular partial derivative can be obtained from the Taylor series expansion of the function with respect to the corresponding variable. For example, the expansion of the function $p(x, t)$ with respect to the coordinate x will have the form (5).

$$p(x_i + \Delta x, t_j) = p(x_i, t_j) + \Delta x \cdot \frac{\partial p}{\partial x} + \ldots \tag{5}$$

If all small terms in expression (5) are discarded, it becomes a right finite difference (6):

$$\begin{aligned} \frac{\partial p}{\partial x} &\approx \frac{p(x_i + \Delta x, t_j) - u(x_i, t_j)}{\Delta x} = \frac{p_{i+1,j} - p_{i,j}}{\Delta x} \\ \frac{\partial p}{\partial t} &\approx \frac{p(x_i, t_j + \Delta t) - u(x_i, t_j)}{\Delta t} = \frac{p_{i,j+1} - p_{i,j}}{\Delta t} \end{aligned} \tag{6}$$

Using left finite differences, one can express the equation as:

$$\begin{aligned} \frac{\partial p}{\partial x} &\approx \frac{p(x_i, t_j) - p(x_i - \Delta x, t_j)}{\Delta x} = \frac{p_{i,j} - p_{i-1,j}}{\Delta x} \\ \frac{\partial p}{\partial t} &\approx \frac{p(x_i, t_j) - p(x_i, t_j - \Delta t)}{\Delta t} = \frac{p_{i,j} - p_{i,j-1}}{\Delta t} \end{aligned} \tag{7}$$

More precisely, partial derivatives can be computed using central finite differences (8).

$$\frac{\partial p}{\partial x} \approx \frac{p(x_i + \Delta x, t_j) - p(x_i - \Delta x, t_j)}{2 \cdot \Delta x} = \frac{p_{i+1,j} - p_{i-1,j}}{2 \cdot \Delta x}$$

$$\frac{\partial p}{\partial t} \approx \frac{p(x_i, t_j + \Delta t) - p(x_i, t_j - \Delta t)}{2 \cdot \Delta t} = \frac{p_{i,j+1} - p_{i,j-1}}{2 \cdot \Delta t} \tag{8}$$

If we apply expression (4) to (1) and then consider expressions (6), (7), and (8), we will be able to numerically solve the partial derivative equation for mass conservation (9):

$$\frac{\partial p}{\partial t} + \frac{\partial\left(p \cdot v_{max} \cdot \left(1 - \frac{p}{p_{max}}\right)\right)}{\partial x} = 0$$

$$\frac{\partial p}{\partial t} + \frac{\partial\left(p \cdot v_{max} - \frac{v_{max} \cdot p^2}{p_{max}}\right)}{\partial x} = 0$$

$$\frac{\partial p}{\partial t} + \left[v_{max} - 2 \cdot \frac{v_{max}}{p_{max}} \cdot p\right] \frac{\partial p}{\partial x} = 0$$

$$\frac{p_{i,j+1} - p_{i,j}}{\Delta t} + \left[v_{max} - 2 \cdot \frac{v_{max}}{p_{max}} \cdot p_{i,j}\right] \cdot \frac{p_{i+1,j} - p_{i-1,j}}{2 \cdot \Delta x} = 0 \tag{9}$$

we express $p_{i,j+1}$ If from (9) given that $p_{i,j} = \frac{p_{i+1,j} - p_{i-1,j}}{2}$, then, from this, we can derive a final expression for calculation (10).

$$p_{i,j+1} = \frac{p_{i+1,j} - p_{i-1,j}}{2} - \frac{\Delta t}{2 \cdot \Delta x} \cdot \left[v_{max} - 2 \cdot \frac{v_{max}}{p_{max}} \cdot \frac{p_{i+1,j} - p_{i-1,j}}{2}\right] \cdot p_{i+1,j} - p_{i-1,j} \tag{10}$$

Depending on how additional conditions are specified to obtain a particular solution of a differential equation, there are two different types of problems [6]:

(a) Initial Value Problem (The Cauchy problem);
(b) Boundary Value Problem.

If these conditions are specified at a single point, such a problem is called an Initial Value Problem (IVP). Additional conditions in the Initial Value Problem are called initial conditions, and the point $x = x_0$ or $t = 0$ (the solution is sought for $t \geq 0$) at which they are specified is called the initial point.

A problem is called a boundary value problem if additional conditions are specified at more than one point, (i.e., at different values of the independent variable). The additional conditions are called boundary conditions. In practice, boundary conditions are usually specified at two points $x = a$ and $x = b$, which are the domain's boundaries.

3 Results

Data on road traffic on the Tashkent Ring Road with the intersection of Bogishamol Street was collected in November 2023. Data is collected every minute and saved in a database. A special video detector is used to detect vehicles and their speeds. The total hourly traffic volume is 570 vehicles per hour during non-peak hours. During peak hours, this figure increases to 1.5–2 times, reaching 870–1065 vehicles per hour.

In this section, the authors will select the numerical method presented in (10) as a test case. The results demonstrate the effectiveness of the current method for the LWR traffic flow model. In all examples, the following speed-density relationships (4) are chosen.

Determining how information about a traffic jam ahead will propagate through the traffic flow is necessary. Solving this problem will make it possible to answer questions (e.g., if a traffic jam forms at the intersection before the traffic lights during rush hour, how long will it take for the jam to reach the previous traffic light?).

The calculation results of the LWR model are presented in the examples below.

Example 1. Calculation parameters: $L = 7000\,m, T = 400\,s, \Delta x = 70\,m, \Delta t = 0.4\,s, v_{max} = 10m/s$. The initial conditions accepted as initial data are as follows: $p_0 = 0, p_L = 0.05\,veh/m, p_R = 0.05veh/m$, and $p_{max} = 0.25\,veh/m$. The solution of the example is presented in Fig. 1.

Example 2. Calculation parameters: $L = 7000\,m, T = 400\,s, \Delta x = 70\,m, \Delta t = 0.4\,s$, and $v_{max} = 16.7m/s$. The initial conditions accepted as initial data are as

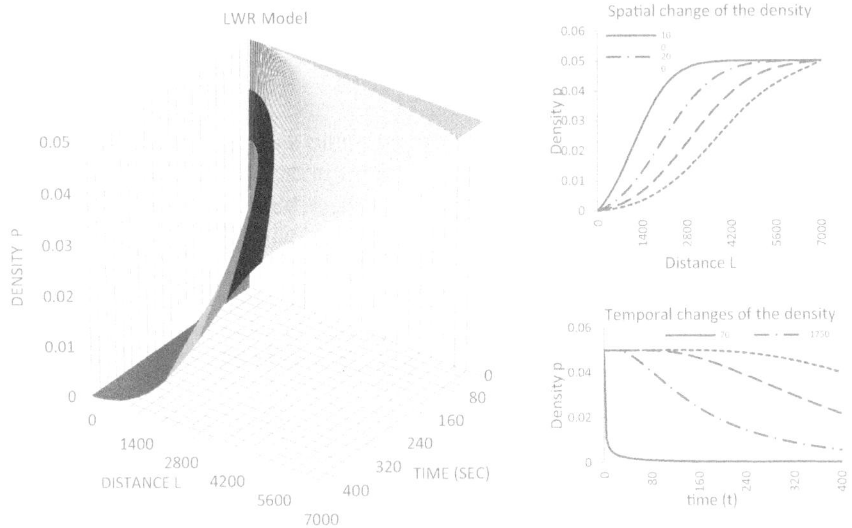

Fig. 1 Solution of the example at a speed of 36 km/h. *Source* Calculated and built by the authors

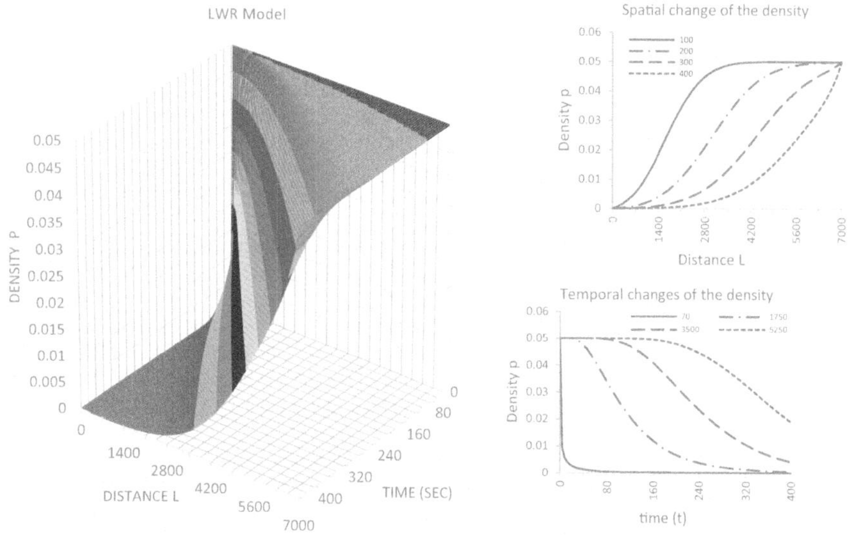

Fig. 2 Solution of the example at a speed of 60 km/h. *Source* Calculated and built by the authors

follows: $p_0 = 0$, $p_L = 0.05\,veh/m$, $p_R = 0.05\,veh/m$, and $p_{max} = 0.25veh/m$. The solution of the example is presented in Fig. 2.

4 Conclusion

The main objective of this research was to investigate the effectiveness of first-order macroscopic models for representing the behavior of traffic flows. For this purpose, the authors selected numerical modeling methods. As a result of this research, the authors found that the higher the speed of traffic movement, the sharper the changes in density from one state to another. From the density graphs with respect to time and space, we can observe that the higher the speed, the more rapidly zero density is approached.

References

1. Chen J, Shi Z, Hu Y (2007) Numerical solution of a two-class LWR traffic flow model by high-resolution central-upwind scheme. In: Shi Y, van Albada GD, Dongarra J, Sloot PMA (eds) Computational science—ICCS 2007: 7th international conference. Beijing, China, pp 17–24. https://doi.org/10.1007/978-3-540-72584-8_3

2. Dubey VP, Kumar D, Alshehri HM, Dubey S, Singh J (2022) Computational analysis of local fractional LWR model occurring in a fractal vehicular traffic flow. Fractal Fract 6(8):426. https://doi.org/10.3390/fractalfract6080426
3. El Ouenjli H, Chafi A, Alami SK (2022) Numerical resolution of the LWR method for first order traffic flow model. In: Motahhir S, Bossoufi B (eds) Digital technologies and applications. Springer, Cham, Switzerland, pp 727–736. https://doi.org/10.1007/978-3-031-02447-4_75
4. Friedrich J, Kolb O, Göttlich S (2018) A Godunov type scheme for a class of LWR traffic flow models with non-local flux. Netw Heterogen Media 13(4):531–547. https://doi.org/10.3934/nhm.2018024
5. Gaddam HK, Chinthireddy A, Rao KR (2016) Comparison of numerical schemes for LWR model under heterogeneous traffic conditions. Period Polytech Transp Eng 44(3):132–140. https://doi.org/10.3311/PPtr.8297
6. Gasnikov AV, Klenov SL, Nurminsky EA, Kholodov YA, Shamray NB (2014) Introduction to mathematical modeling of traffic flows: textbook. MCNMO University, Moscow, Russia
7. Goatin P, Scialanga S (2016) Well-posedness and finite volume approximations of the LWR traffic flow model with non-local velocity. Netw Heterogen Media 11(1):107–121. https://doi.org/10.3934/nhm.2016.11.107
8. Holden H, Risebro NH (2017) Follow-the-Leader models can be viewed as a numerical approximation to the Lighthill-Whitham-Richards model for traffic flow. Netw Heterogen Media 13(3):409–421. https://doi.org/10.3934/nhm.2018018
9. Imran W, Khan ZH, Gulliver TA, Khattak KS, Saeed S, Aslam MS (2021) Macroscopic traffic flow characterization for stimuli based on driver reaction. Civil Eng J 7(1):1–13. https://doi.org/10.28991/cej-2021-03091632
10. Khan MU, Saeed S, Nehdi ML, Rehan R (2021) Macroscopic traffic-flow modelling based on gap-filling behavior of heterogeneous traffic. Appl Sci 11(9):4278. https://doi.org/10.3390/app11094278
11. May AD (1990) Fundamentals of traffic flow. Prentice Hall, Englewood Cliff, NJ
12. Ni D (2015) Traffic flow theory: characteristics, experimental methods, and numerical techniques. Butterworth-Heinemann, Oxford, UK. https://doi.org/10.1016/C2015-0-01702-6
13. Polson N, Sokolov V (2015) Bayesian analysis of traffic flow on interstate I-55: The LWR model. Ann Appl Stat 9(4):1864–1888. https://doi.org/10.1214/15-AOAS853
14. Qin Y, Wang H, Ni D (2021) Lighthill-Whitham-Richards model for traffic flow mixed with cooperative adaptive cruise control vehicles. Transp Sci 55(4):883–907. https://doi.org/10.1287/trsc.2021.1057
15. Setiawan EB, Tarwidi D, Umbara RF (2015) Numerical simulation of traffic flow via fluid dynamics approach. Int J Comput Optim 3(1):93–104. https://doi.org/10.12988/ijco.2016.6613
16. Treiber M, Kesting A (2013) Traffic flow dynamics: data, models and simulation: textbook. Springer, Berlin, Heidelberg, Germany, pp 983–1000. https://doi.org/10.1007/978-3-642-32460-4
17. Yuan Y, Duret A, Van Lint H (2015) Mesoscopic traffic state estimation based on a variational formulation of the LWR model in Lagrangian-space coordinates and Kalman filter. Transp Res Procedia 10:82–92. https://doi.org/10.1016/j.trpro.2015.09.058

The Importance of Transportation Infrastructure in the Regional Economic Development

Komil B. Urazov and Zumrad A. Orzukulova

Abstract The research explores the impact of transportation use and regional transport infrastructure on enhancing the efficiency of transport and logistics services in Uzbekistan. During the research, the authors conducted a comparative analysis of the coefficients representing the density of the transport network in the regions for 2021 and 2022. For 2022, the authors developed regression equations to represent the influence of regional transport infrastructure indicators on socio-economic outcomes in Uzbekistan. The analysis examined the effects of these factors, revealing that Andijan, Fergana, and Namangan have high transport network density, while the Republic of Karakalpakstan, Navoi, and Bukhara have lower density. The correlation analysis shows that a 1% change in the number of cars, the length of paved roads, and the length of public railways in a region results in more than a 0.5% change in the region's GRP. Therefore, enhancing transport infrastructure indicators can further stimulate regional economic development.

Keywords Transport networks · Transport use · Transport infrastructure · Transport-logistics services · Transport supply

JEL codes R4 · O18 · R420

1 Introduction

The level of regional economic development is largely related to the availability and level of transport provision. Transport infrastructure development is important for local, regional, and national security. The transport system carries out such functions as transportation of goods and passengers, provision of economic connection

---

K. B. Urazov
Samarkand Institute of Economics and Service, Samarkand, Uzbekistan
e-mail: komil.urazov@inbox.ru

Z. A. Orzukulova (✉)
Social and Political Sciences Institute, Tashkent, Uzbekistan
e-mail: zumrad.orzuqulova@mail.ru

S. M. Sultanova et al. (eds.), *Sustainable Development of Transport*, Lecture Notes in Networks and Systems 1366, https://doi.org/10.1007/978-3-031-88846-5_11

of regions, provision of employment of the population, and reduction of variable costs. The number of vehicles used in transportation and storage services, along with their regional distribution and availability, are essential indicators for assessing the efficiency of transport and logistics services.

Globally, well-organized transport infrastructure and high transport availability are key factors in establishing and successfully operating modern cluster centers. Cluster centers are commonly found in high-income developed countries. In such countries, significant attention is paid to the transport infrastructure. Historically, large, advanced, multimodal logistics clusters (e.g., integrated logistics centers, land ports, shipping villages, and extended port complexes) have emerged in high-income countries. Particularly, the USA has the most efficient, largest, and best-connected cluster center with the necessary infrastructures in the world. Singapore has the largest shipping hub port in the world. Western and Central European countries are famous for the number of cluster centers per land area. Logistics cluster centers are aimed at ensuring competitiveness by increasing the efficiency of logistics services. More than half of the top 30 countries with the best results in the World Bank's logistics efficiency index are European countries; except China, all countries in the top are high-income countries. All of these countries have developed logistics clusters. It is desirable to establish logistics clusters in countries with middle incomes and geographical advantages for multimodal transportation. However, ensuring that these logistics clusters operate as interconnected networks is essential. Such cluster centers are rare in low-income countries.

It would be appropriate for Uzbekistan to use this experience to improve the efficiency of transport logistics services, improve its position in the international rating, and participate in other international ratings. To organize such logistics clusters, it is necessary to create an infrastructure and the opportunity to operate as a network. Primarily, it is necessary to pay attention to the transport infrastructure, the potential of the regions, and sources of financing.

2 Materials and Methods

Various studies consider the impact of transport infrastructure on the region's economy, calculate the level of transport provision, focus on the impact of economic development on the transport system, and propose improving transport infrastructure using advanced technologies and mobile applications. Ali et al. [1]; Beuthe et al. [2]; Blancas [3]; Camisón-Haba and Clemente-Almendros [4]; Karpenko and Karpenko [6]; Kataeva [7]; Lai [8]; Magomedov [9]; Notchenko and Lineva [10]; Polyakov and Nikulina [11]; Seliverstov and Seliverstov [12]; Tsegelnyuk [14]; and Vakhrameev [15]. Moreover, the considered issues are discussed in the work of Ejiogu et al. [5].

Foreign scholars consider the level of transportation as a quantitative indicator. There are several methods for determining it. For instance, Kataeva proposed creating a ranking of regions using an integrated assessment that includes six groups of indicators [7]. Correlation analysis showed the relationship between technical indicators and economic indicators. According to the methodology proposed by Seliverstov [12], key indicators (e.g., freight turnover, road length, transferability, vehicle availability, passenger turnover, and transport income and expenses) were included in the analysis to more comprehensively assess the level of transport infrastructure and its capacity.

The concepts of transport availability and the provision of transport infrastructure are closely intertwined with the social and economic development of a region. There are the following economic indicators:

- Standard of living of the population;
- Investment volume;
- Level of employment;
- Freight turnover;
- Volume of transported cargo;
- Gross regional product, and others.

The authors define the concepts of availability of transport and provision of transport infrastructure.

Transportation availability represents the degree of connectivity of an area by representing the ability of people to use the transportation system to get to any point in the country.

The availability of transport means the possibility of using transport services within the level of development of the existing transport infrastructure. This indicator consists of technological and monetary components. The technological component includes the availability and readiness to transport cargo in various conditions at any time. The monetary component is the alternative to prices for transport services in terms of price.

These two indicators strengthen local and international relations, increase the volume of cargo turnover, and create favorable conditions for organizing an optimal logistics system.

Factors such as proximity to neighboring countries, tourism accessibility, industrialization, and the availability of railway crossings and ports also influence the development of the transport system. Currently, no standardized indicator exists to measure the level of transport system development, as regions vary significantly in their social, economic, natural, and geographic characteristics.

The coefficients of Engel, Goltz, Uspensky, and Vasilevsky are widely used. These coefficients consider the density of traffic calculated per 1000 km^2 and 10,000 people. The authors also use these coefficients to divide regions into groups and determine their potential. The coefficients are as follows:

1. Engel coefficient: $d_{LSH} = \frac{L_E}{\sqrt{SH}}$;
2. Density of transport network per 1000 km^2: $p_1 = \frac{L_E \bullet 1000}{S}$;
3. The density of the transport network per 10,000 people: $p_2 = \frac{L_E \bullet 10000}{H}$;
4. Holtz coefficient: $K_g = \frac{L_E}{\sqrt{S \bullet P}}$;
5. Uspensky coefficient: $K_u = \frac{L_E}{\sqrt[3]{S \bullet H \bullet t}}$.

Based on these formulas, the authors calculate the coefficients for 2021 and 2022.

For calculation, the authors used data from the statistical bulletin for 2019–2022 obtained from the Statistics Agency under the President of the Republic of Uzbekistan [13]. For the primary data, the authors selected indicators such as area, population, length of public railways and highways, number of settlements, and the volume of cargo transported by rail and road. In the formulas, measurement units were initially in their smallest denominations (km, person, and kg). For clarity in the tables, the authors scaled them to thousand kilometers, thousand people, and tons. This adjustment does not affect the mathematical results. The authors calculated the relevant coefficients for railway and road transport. Road transport accounts for a significant portion of cargo turnover, while rail transport also holds a considerable share. The authors did not include air transport due to its relatively small share. Additionally, the authors did not consider pipeline transport because it requires a separate infrastructure and was outside the scope of this analysis.

The density of transport networks per 1000 km2 on the territory of Uzbekistan did not change significantly in 2021 and 2022. Andijan, Fergana, and Namangan Regions are the regions with high density of transport networks. Low-density areas include Karakalpakstan, Navoi, and Bukhara.

In terms of the density of transport networks per 10,000 people, a change of 0.01%–0.03% was observed in the regions from 2022 compared to 2021. In 2022, Navoi, Karakalpakstan, and Bukhara Regions can be seen as regions with high density. Andijan, Fergana, and Samarkand Regions are low-density areas.

The changes from 2021 to 2022 in the Engel, Goltz, and Uspensky coefficients were minimal. Figure 1 compares the regions by these three coefficients. In this bar chart, the Engel and Uspensky coefficients are very close to 0, almost overlapping.

Regions with a high Goltz coefficient include Andijan, Namangan, Syrdarya, and Fergana, while the Republic of Karakalpakstan, Bukhara, and Navoi have lower values. Similarly, Fergana, Syrdarya, and Namangan lead in the Engel coefficient, consistent with the other two coefficients, whereas the Republic of Karakalpakstan, Kashkadarya, and Navoi show the lowest results.

Namangan and Ferghana recorded the highest result in the Uspensky coefficient; Navoi recorded the lowest result.

Thus, Namangan and Fergana show strong results across all three analysis methods. In contrast, the Republic of Karakalpakstan and Navoi have lower performance, primarily due to their low population density, as much of these regions consist of desert areas. However, this situation is expected to improve as desert zones are further developed.

As already mentioned, the general condition of the transport network cannot be expressed only by these indicators. The Goltz and Uspensky coefficients are modified versions of the Engel coefficient, designed to provide a more comprehensive assessment of the state of the transport network.

Following the approach of Yu. V. Kataeva, the authors will begin to theoretically and econometrically justify the connection between well-developed transport infrastructure and the economic performance of the regions.

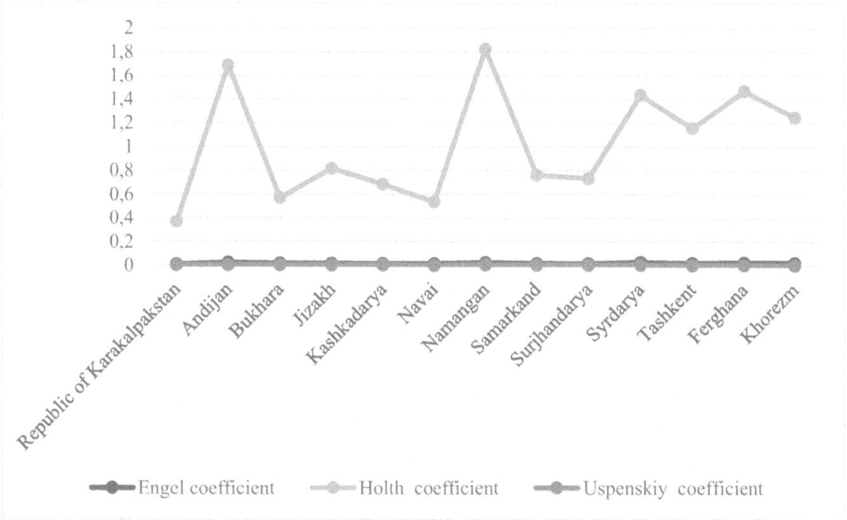

Fig. 1 Coefficients of the density of transport networks of the republic's territories in 2022. *Source* Calculated and compiled by the authors based on [13]

From a theoretical point of view, improving the region's transport infrastructure affects the volume of passenger and cargo transportation, jobs, industrial enterprises, local budget revenues, and expenses. In turn, the development of the region's economy also affects the state of the transport infrastructure. The presence and improvement of transport infrastructure significantly affect a region's economy by accelerating processes, increasing production opportunities, reducing accidents on roads built to high-quality technical standards, and minimizing vehicle wear and tear. However, it is important to note that transport networks can also harm the local environment. As a result, efforts are being made to promote the use of environmentally friendly fuels. While the benefits of improved transport infrastructure are clear, such development is closely linked to the region's economy. The economic growth of a region drives the construction of new transport facilities, the reconstruction of existing ones, the mobilization of labor, and the supply of material and technical resources. Local authorities manage and regulate traffic flow while attracting domestic and foreign investment in transport infrastructure.

The authors substantiate the theoretical representation of the relationship between regional economies and transport infrastructure through correlational analysis. The primary objective of this analysis is to assess the strength of the relationship between direct and indirect indicators of transport infrastructure activity and development, as well as the socio-economic indicators of the region. The authors' approach differs from that of Yu. V. Kataeva, because they focus specifically on transport infrastructure related to cargo transportation and storage, excluding passenger transportation. This research centers on studying cargo transportation and the factors that influence it.

The authors used the indicators representing the socio-economic development and transport infrastructure indicators of the region for 2022. These group indicators serve as the initial point for our econometric analysis.

Using this set of indicators, the authors established the correlation relationship by utilizing the "Correlation" function in the Data Analysis section of Microsoft Excel 2013. As a result, the authors analyzed the influence of factors (transport infrastructure indicators) on each variable (socio-economic indicators) individually. By focusing on the first columns of the resulting correlation matrices, the authors identified the correlation between the two indicators (Table 1).

From this correlation, it is evident that the two factors with the greatest influence are Y1-GDP and Y2-Number of Operating Transportation and Storage Enterprises. These two variables form the basis of our model. Air, water, and pipeline transport are excluded from the model because the statistical share of water and air transport is significantly smaller, and regional data for pipeline transport is unavailable. Now, the authors will construct an equation representing the two regression relationships.

$$\begin{aligned} Y_1 = & -5051 + 122.6X_1 + 350\,X_2 + 1020.8\,X_3 + 109.1\,X_4 \\ & -3X_5 - 3.3X_6 + 11.4\,X_7 - 14.3\,X_8 - 441.3\,X_9 \end{aligned} \quad (1)$$

$$\begin{aligned} Y_2 = & -576.2 + 2.4X_1 + 5\,X_2 - 33.3X_3 - 1.7X_4 \\ & + 1.2X_5 - X_6 + 0.5\,X_7 + 1.8\,X_8 + 29.8\,X_9 \end{aligned} \quad (2)$$

In these models, indicators X1, X4, X5, and X9 are relatively significant in the first model according to Student's criterion. These indicators are the number of cars, car freight, car traffic turnover, and number of accidents. In fact, since the authors have a much larger share of transportation by car, these factors affect the size of the region's GRP and the number of transportation and storage enterprises in the region.

3 Results

The economic significance of the obtained results and the elasticity indicators of the factors are presented in Table 2, which shows how the volume of GDP and the number of transport and storage enterprises change when each factor changes by 1% or 1 unit.

Analyzing the elasticity indicators from the table, high elasticity was demonstrated by factors X1, X7, X8, and X9 in the first model and factors X1, X5, X6, X7, and X8 in the second model. This is due to the fact that a 1% change in the number of cars, the length of paved roads, and the length of public railways results in more than a 0.5% change in GNI.

If we forecast using this model, we will be able to determine the size of the GNI by calculating a 1% increase in the number of cars, the length of paved roads, and the length of public railways for 2024–2025. Data for 2022 were calculated in the work.

Table 1 Correlational basis of the relationship between the transport infrastructure and socio-economic indicators of the regions

Indicators of transport infrastructure development		GRP by transportation and storage, communication and information activities, billion soums	Overall GRP of the region, billion soums	Population, thousand people	Number of employed in the economy, thousand people	Income per capita, thousand soums	Volume of industrial products, billion soums	Volume of agricultural products, billion soums	Number of operating transportation and storage enterprises, pcs
	Number of cars, thousand units	0.68	1.00	1.00	1.00	0.05	0.24	0.87	1.00
	Number of trucks, thousand units	0.68	0.99	1.00	1.00	0.03	0.12	0.81	0.99
	Freight sent by railway, million tons	0.56	0.97	0.94	0.94	0.24	0.77	0.00	0.96
	Transportation of goods by road, million tons	0.59	0.97	0.94	0.94	0.30	0.86	−0.13	0.95
	Road transport freight turnover, million t-km	0.64	1.00	0.99	0.99	0.13	0.58	0.47	1.00
	Length of paved roadways, thousand km	0.65	1.00	0.99	0.99	0.08	0.41	0.33	0.99

(continued)

Table 1 (continued)

	Length of improved paved highways, thousand km	0.65	1.00	0.99	0.99	0.08	0.40	0.29	0.99
	The length of railways in general use km,	0.61	0.98	0.98	0.98	0.09	0.07	−0.29	0.98
	Number of car accidents, per 100,000 population	0.10	0.10	0.07	0.08	0.01	0.59	0.11	0.11

Source Calculated and compiled by the authors based on [13]

Table 2 Economic description of the econometric model showing the impact of transport infrastructure on the regional economy

The rise of indicator to 1% or 1 unit	The growth (fall) of GRP (Y1)		The growth (fall) of the number of active transportation and storage enterprises (Y2)	
	Billion sum	%	Pieces	%
X1—Number of cars, thousand units	107,413.4	0.510	2201.4	0.525
X2—Number of trucks, thousand units	107,121.5	0.097	2195.4	0.072
X3—Freight sent by railway, million tons	107,151.8	0.100	2195.7	−0.170
X4—Transportation of goods by road, million tons	107,129.1	0.011	2195.9	−0.142
X5—Road transport freight turnover, million t-km	107,167.9	−0.005	2242.6	1.533
X6—Length of paved roadways, thousand km	108,731.4	−0.084	2375.2	−2.964
X7—Length of improved paved highways, thousand km	107,558.3	0.644	2243.5	1.533
X8—The length of railways in general use km	107,202.8	−0.787	2202.3	0.583
X9—Number of car accidents, per 100,000 population	107,432.7	−2.781	2192.7	0.383

Source Calculated and compiled by the authors

The year 2022 is taken as the base year. In this case, the forecast for the number of cars, length of improved paved highways, and length of railways in general use accounted for 1–3%. The volume of GRP will be 1,341,396 billion soums in 2024; in 2025, it will reach up to 2,025,508 billion soums.

In the forecast data, we received projections not for each region individually but for the entire republic. It shows that if the number of cars increases by 1% to 3,430,645 units in 2024, the GNP is expected to reach 1,341,396 billion sums, driven by the growth of transport services in the region.

4 Conclusion

Thus, using the example of automobiles and railways, the authors observed that improvements in vehicles and the infrastructure they rely on contribute to the growth of GNP and promote the socio-economic development of the region. Uzbekistan can boost transit freight volumes by enhancing its road and railway infrastructure. This will be a key factor in driving economic development and increasing the region's investment attractiveness.

References

1. Ali Y, Rahman A, Lala S, Sabir M (2021) Economic viability of foreign investment in public transport of Pakistan—orange line metro train in focus. Trans Corp Rev 13(3):321–333. https://doi.org/10.1080/19186444.2020.1851637
2. Beuthe M, Gasca J, Greene D, Lee DS, Muromachi Y, Newton PJ, Zhou PJ et al (2007) Transport and its infrastructure. In: Netz B, Davidson OR, Bosch PR, Dave R, Meyer LA (eds) Climate change 2007: mitigation. contribution of working group III to the fourth assessment report of the intergovernmental panel on climate change. Cambridge University Press, Cambridge, UK; New York, NY, pp 324–385. https://www.ipcc.ch/site/assets/uploads/2018/02/ar4-wg3-chapter5-1.pdf. Accessed 30 Apr 2024
3. Blancas LC (2022) Logistics clusters in the era of disruption. World Bank Blogs. https://blogs.worldbank.org/en/transport/logistics-clusters-era-disruption. Accessed 30 Apr 2024
4. Camisón-Haba S, Clemente-Almendros JA (2019) A global model for the estimation of transport costs. Econ Res 33(1):2075–2100. https://doi.org/10.1080/1331677X.2019.1584044
5. Ejiogu EO, Madonsela NS, Adetunla A (2020) The effect of transportation infrastructure on economic development. In: Proceedings of the 2nd African international conference on industrial engineering and operations management. IEOM Society, Harare, Zimbabwe
6. Karpenko EM, Karpenko VM (2020) Criteria for assessing the efficiency of functioning regional transportation and logistics system. Proc BSTU Econ Manag 2(238/5):84–90
7. Kataeva YuV (2013) Integrated estimation of the level of regional transport infrastructure development. Perm Univ Herald Econ 4(19):66–73
8. Lai Q (2020) The transportation infrastructure and regional economic growth—evidence from Dongguan Humen bridge. Mod Econ 11(12):2055–2080. https://doi.org/10.4236/me.2020.1112137
9. Magomedov AM (2013) Transport subsystem in logistics systems region. Manag Econ Syst Electr Sci J 11(59):15
10. Notchenko VV, Lineva EL (2015) Assessment of transport infrastructure of a region as a necessary condition for entrepreneurship development. St. Petersburg State Polytech Univ J Econ 1(211):172–181. https://doi.org/10.5862/JE.211.19
11. Polyakov KV, Nikulina SV (2016) Strategy for the development of the transport system of St. Petersburg and the Leningrad region until 2030. Transp Russian Fed 5(66):36–41
12. Seliverstov YA, Seliverstov SA (2021) Development of a quality analysis system for street and road networks based on combined features. Transp Sci Equip Manag 8:14–29. https://doi.org/10.36535/0236-1914-2021-08-3
13. Statistics Agency under the President of the Republic of Uzbekistan (2023) Transport and communications in Uzbekistan. Tashkent, Uzbekistan. https://stat.uz/uz/nashrlar/3671-o-zbekistond-tr-nsport-v-loq. Accessed 30 Apr 2024

14. Tsegelnyuk VA (2016) Estimation of transport factors influence on the economy of the far eastern capital territories. Vestnik Astrakhan State Tech Univ (Series: Economics) 4:123–132
15. Vakhrameev II (2014) Influence of transport infrastructure on the branch development of the region's economy. Bull Transbaikal State Univ 08(111):85–91

Implementation of Artificial Intelligence Tools (Smart City Technologies) for Managing Infrastructure in Urbanized Systems

Alexandra V. Gridchina, Yulia V. Aleksakhina, Elena A. Gorshkova, and Georgii V. Gridchin

Abstract The research aims to describe new approaches and directions in the use and development of smart city technologies. The authors apply comparative analysis to identify the correlation between digitalization (smart city technologies) and the improvement of people's quality of life in cities, using Moscow as an example. The research object is the digital infrastructure of Moscow. The prospects and realities of AI as one of the main drivers of smart city development are undeniable. However, rapid progress has also revealed various moral and ethical issues. Without addressing these, the successful creation of smart cities cannot be assured. Developing ethical principles and behavioral standards for using AI systems is necessary. Based on the analysis of changes in the quality of life in large metropolises, the authors determined that a new quality of life and economic growth has emerged in cities under the influence of the smart economy. The advantages of digitalization include increased purchasing power, advances in healthcare, reduced inflation, and less commuting time due to the growth of online economic activities. However, the issue of adhering to ethical standards is highly relevant in the smart economy; its resolution will impact the further improvement of the quality of life in cities.

Keywords Artificial intelligence · Smart city · Digital economy · Control center · Smart economy · Digital infrastructure

JEL codes O32 · O35 · O38 · Q55

A. V. Gridchina (✉) · Y. V. Aleksakhina
Moscow Polytechnic University, Moscow, Russia
e-mail: gav70@bk.ru

E. A. Gorshkova
Prince Alexander, Nevsky Military University" of the Ministry of Defense of the Russian Federation, Moscow, Russia

G. V. Gridchin
State University of Management, Moscow, Russia

S. M. Sultanova et al. (eds.), *Sustainable Development of Transport*, Lecture Notes in Networks and Systems 1366, https://doi.org/10.1007/978-3-031-88846-5_12

1 Introduction

The concept of a smart city (i.e., the digitalization of urban infrastructure) is now widely recognized due to its practical implementation. It often reflects many key aspects of quality of life. The integration of information and communication technologies in the creation of smart urban systems undeniably impacts the quality and satisfaction of life: digital transformation signifies diversity, the efficiency of urban services and operations, the quality and speed of their delivery, transaction security, personal safety, time savings, and overall optimization. A smart city significantly enhances our lives through AI, the Internet of Things, autonomous transport, big data, blockchain, cloud technologies, augmented reality, and 5G. It represents a fusion of AI, finance, and urban environments.

2 Materials and Methods

The specifics and prospects of forming a smart economy in cities are explored in the works cited in this research.

Despite many publications on related topics, the issue of the qualitative transformation of economic growth in the context of a smart economy and the potential benefits for improving people's quality of life has not been sufficiently studied or resolved in contemporary economic science.

A smart city is a part of the country's digital economy. In essence, it involves information and communication technologies to manage the city—its assets, transportation, water supply, and energy, among other areas. It aims to use resources efficiently and improve the lives of citizens.

Here are just a few specific examples of the impact of implementing smart city technologies on the daily lives of urban residents:

- A 10–30% increase in indicators of quality of life and resident satisfaction;
- 15–30 min of time saved on daily commutes using public transport;
- An 8–15% reduction in illness rates;
- A 1–3% decrease in the cost of living;
- A 25% increase in social connectivity and civic engagement with local authorities;
- A reduction in water consumption by 25–80 L per person;
- A 1–3% increase in the number of jobs;
- A 15% reduction in the costs of creating and maintaining infrastructure [8].

The urban population is growing worldwide, with vast megacities inhabited by millions becoming the norm. However, managing such urban systems is no longer possible with traditional analog methods. As a result, cities of all sizes are rapidly becoming smart, incorporating artificial intelligence.

While only 14% of the world's population lived in cities in 1900, by 2010, half of the global population was urbanized. This trend continues. In 1950, there were about 80 cities with populations over a million, whereas by 2018, this number had risen to 500. Russia is no exception. As of the beginning of 2024, over 75% of the country's total population lives in cities [11].

The infrastructure of a smart city is a high-tech system composed of numerous interconnected elements—computers linked into a unified network, controllers, sensors, and devices generating vast amounts of data. This data must be accumulated, stored, transmitted, analyzed, and processed during and after its collection because it will be needed to build databases and data banks for future administrative tasks. This means that a highly advanced infrastructure is required for smart technologies to function in a city. Moreover, tasks must be prioritized according to certain priorities [10].

In the world of cutting-edge technology, the concept of a smart city is no longer new. Many residents of megacities have become accustomed to the consumer-oriented improvements that today's cities offer. The ease of accessing information, the rapid response to requests, and the elimination of bureaucratic paperwork have already become the norm for urban dwellers. However, this is the perspective of the city from the citizen's viewpoint.

Artificial intelligence has become a key element of the smart city. The reality is that AI is a powerful driver of cities' digital development.

The growth of the urban population caused the expansion of transportation infrastructure. The need for smart systems became apparent. To ensure convenience, safety, and accessibility for urban mobility, it is essential to develop a "Smart Transportation Infrastructure"—a combination of computer vision and machine learning for the autonomous movement of all types of transport. For instance, a unified city transportation system operating in real-time can factor in data on movement, optimization scenarios, and travel speeds, thereby improving safety and quality of life for citizens.

As cities have grown, the demand for various municipal services (e.g., healthcare, social services, and utilities) has also increased. The use of computing technology in administrative activities (in terms of speed and labor intensity) makes a city's economy as automated as possible, turning the city into a smart one. Naturally, the first step toward making a city smarter is data collection. According to recent data, large cities and metropolitan areas generate and handle petabytes of data per month, and such volumes cannot be managed without AI. AI can employ various analytical methods and predictive analytics, making it possible to foresee emerging issues. Assisting in data collection and processing is a complex but solvable task for AI.

Sources of information include various sensors, data on purchases and sales, different types of statistics, reports (from municipal services, healthcare services, law enforcement agencies, results of environmental monitoring, and more), videos, and photos. Image recognition technologies make the use of AI possible [9].

AI is becoming a key tool for managing and coordinating smart cities. It allows for processing vast amounts of data, turning it into valuable information, delivering convenient and useful services to citizens, improving infrastructure, and increasing the efficiency of urban systems. Undoubtedly, AI plays an increasingly significant role in shaping and developing urban environments, ensuring comfort, safety, and sustainable development for residents.

It is important to note that these benefits are just a few of the many possibilities that AI tools offer for the functioning of urban infrastructure. The potential for significantly improving the quality of life for citizens exists and must be studied and integrated into practice. In the future, the use of AI in urban management will be seen as an essential tool for increasing the efficiency of city governance.

Thus, let us delve into the current practices of using AI in managing urban systems.

While city management was previously done mostly manually, almost all processes are currently digitized. The areas where AI is applied in urban management are constantly expanding. Any city's urban system is filled with enormous amounts of data, necessitating the presence of information systems.

If we look at the urban infrastructure of a megacity like Moscow, the following data can be provided. The city operates over 70,000 buildings and structures, 16,000 km of heating networks (which exceeds the distance from Moscow to the South Pole), 8000 km of drainage collectors (equal to the length of the road from Moscow to Khabarovsk), and 80,000 km of underground power grids (equivalent to twice the circumference of the Earth along the equator). Providing communications, energy, water, and heat for residents and businesses, managing public amenities and capital repairs, utilities, and personnel management are complex tasks that require maximum coordination. More than 50 organizations within Moscow's urban management complex have already developed and implemented digital information systems that cover various aspects of their activities.

It is essential to centralize information flows to effectively manage urban infrastructure. Digitalization and centralization of management are key trends in the development of smart cities, realized through a unified digital control center based on AI technologies. The following areas of development can be highlighted: the digital transformation of urban infrastructure and its management, ensuring citizen participation, public interaction, and the digitalization of urban planning policies.

The Urban Infrastructure Management Center (UIMC) has been created in Moscow to oversee the functioning of more than a thousand parameters of urban life. Information flows into the center from over 100 sources, including budgetary institutions and government bodies. Data is received automatically (through software interfaces), semi-automatically (via email), and manually (by filling out special forms). Each organization involved in urban development, utilities, ecology, energy, road management, and industry has its own regulations, level of digitalization, and performance indicators. Processing such a diverse and vast amount of information, analyzing it, and making management decisions is impossible without AI.

AI must be trained to be used in urban management. Machine learning requires a large volume of verified data on decisions that lead to desired outcomes. Information is accumulated from various sources, entered into a unified information space, and then analyzed and adapted for easy comprehension (using graphs and charts).

The control center collects data from various sources and presents it in a format that is easy to interpret. However, it is crucial to remember that graphs and charts may limit in-depth analysis of cause-and-effect relationships, making it harder to make well-founded management decisions. Decisions requiring verification across all parameters should only be made by experienced specialists in the field, who are just beginning to emerge. Before creating the UIMC, no one had experience working with such a diverse range of real-time data about urban life. The ability for objective control was extremely limited.

The creation of the UIMC has brought numerous new opportunities for conducting objective oversight of various aspects of urban life. Now, with the help of new technologies and data synchronization from multiple information sources, it has become possible to access a wide range of data on the state of urban infrastructure, the environmental situation, the transportation system, and other aspects of city life.

In today's reality, large cities have digital twins, which enable highly efficient management of urban systems. The history of integrating the latest technologies into urban management began at the turn of the twenty-first century when elements of smart systems started to be applied in urban structures. Nowadays, we are witnessing a new phase in urban development—active implementation of AI technology in the management systems of megacities. Neural networks are used to analyze vast amounts of data to identify issues, make decisions to resolve them, and, essentially, create digital twins of real cities.

One of the latest trends in data collection is the use of satellite data, which enhances the relevance of information gathered in real time and ultimately increases the efficiency of urban infrastructure (an example is real-time data on parking occupancy) [5].

The UIMC enables the monitoring and analysis of collected data, which, in turn, allows for more informed and effective decisions aimed at improving the quality of life for citizens and optimizing the operation of urban services. For instance, the synchronization of waste disposal data allows for more precise control over schedule adherence and punishing violators, similar to how traffic fines are imposed by the State Road Safety Inspectorate.

Thus, the creation of the UIMC has been a significant step in the development of cities, providing access to objective information and enabling well-founded decisions to improve the quality of life for urban residents.

The people who continued their professional activities at the UIMC previously represented various urban organizations and worked in industry, energy, government, or business. Each brought their unique experience and perspective on the challenges and problems the city faces.

In the past, each organization was responsible for specific aspects of city life. For instance, if someone got stuck in an elevator, a specialized organization would handle it; if there was a power outage, another organization would take care of it; if there was a gas leak, another organization would be responsible.

However, with the creation of the UIMC, these organizations were unified into a single structure. This consolidation has reduced response times to various incidents and made it possible to solve problems effectively, as representatives from different sectors can now work together and share information.

Nowadays, when an incident or call occurs, the UIMC staff can immediately take action, coordinating the work of various organizations in real time. This significantly improves the efficiency of responses and minimizes reaction times.

Currently, AI is successfully applied and has proven its worth in Moscow. According to the latest data, more than 70 urban projects are utilizing AI. One of the priorities of urban structures is to create a safe and comfortable environment; AI is indispensable in this regard. One example is the Face Pay project in the Moscow Metro, which optimizes the time it takes to enter the subway and enhances passenger security. Residents are already accustomed to voice assistants, contact centers, and chatbots facilitating communication and interaction with city authorities and services at various levels [3].

One AI-powered project is the smart utilities system. It ensures the operation of various sensors and controllers installed on equipment to transmit readings for subsequent processing, analysis, and data accumulation. The main goal of the smart utilities system is to ensure the safe and uninterrupted operation of the complex, multi-level systems of a city.

Digitalizing security is another promising area of AI application in urban management. Traffic control systems, vehicle monitoring, and satellite navigation are all tools that help address transportation issues and enhance road safety.

However, video camera data is used not only to solve traffic problems. In Moscow, a neural network processes footage from cameras to identify problem areas in the city: trash buildup, snowdrifts, ice hazards, lighting failures, and landscaping issues.

Studying global practices of AI applications in the life support systems of modern cities provides valuable insights and allows for the adaptation of best practices domestically. A widely recognized example is Alphabet Sidewalk Labs, which successfully operates in Toronto. This system monitors and optimizes urban traffic and ensures efficient energy savings [4].

3 Results

Thus, creating the UIMC makes it possible to unify various sectors and tested problem-solving strategies to enhance the quality of life for urban residents. It facilitates more efficient and coordinated management of city resources while improving the safety and comfort of the urban environment.

In addition to understanding related regulations and criticality levels, a staff member of the management center must possess innovative thinking. They must be able to foresee the interconnections between different events and their mutual impact. For example, multiple incidents involving gas, electricity, and visitors trapped in an elevator could occur simultaneously in the same building. In such cases, it is necessary to automatically prioritize incidents, bypassing the regulatory organizations responsible for addressing them.

To effectively apply AI in practice, particularly in management, it is important that specialists in this field make decisions, learn from their mistakes, and make decisions again. This cyclical process will eventually create high-quality training materials for future learning. Such materials are currently scarce because people still tend to perceive AI more as a new toy, including those working in management centers.

For AI to become a real and useful tool in management, time is needed to train and educate the staff working in management centers, government bodies, and public organizations. In addition to time, these individuals must make bold and responsible decisions, be patient, and stay focused on achieving their goals.

A successful real-world case by global standards is the city of Moscow, a smart city (ranked 14th in the world in the Smart City Government Rankings). Ten criteria were used to evaluate smart cities: budget, the presence of innovations, leadership, talent development, regulatory environment, human-centeredness, ecosystems, support programs, development strategy, and finances. In April 2024, Moscow won the grand prize for the second time at the annual National Award for Contributions to the Development of Urban Digitalization, "Smart City" (the award has been presented since 2023) [6].

4 Conclusion

Contemporary directions in smart city development include the following:

- Safety (public, transportation, environmental, and utility safety and coordination during emergencies);
- Digital governance (planning, efficiency of public services, coordination and synchronization, open government, and monitoring);
- Residents' well-being (social policy, lifestyle, and leisure organization);
- The investment environment (sustainable growth and innovation-driven development).

The current reality is that the architecture of a smart city has already been defined. The four levels utilized by consumers are services, industry platforms, data, and digital infrastructure. A digital city has been created around and for people, with all the advantages achieved through the diligent collaboration of human and artificial intelligence: the prioritization of electronic document management, 24/7 online data monitoring, access to key city facilities, AI-driven solutions to urban problems, the parallel existence of the city's digital twin, collaborative development of business and the scientific community, a barrier-free environment, and the participation of citizens in city management. The number of smart cities is also growing. According to MarketsandMarkets, there will be 800 smart cities globally by 2030. Simultaneously, the urban population is expected to increase, with the UN forecasting that by 2050, 70% of the world's population will live in cities [12].

AI is becoming a part of our lives. It is successfully used in the creation and functioning of smart cities, including the following aspects:

- Advanced AI methods (generative AI and high-level vehicle autonomy);
- Natural language processing (use of chatbots in citizen communication);

- Speech recognition and synthesis (use of robot operators for citizen interactions);
- Intelligent decision support (AI in monitoring and forecasting urban management, the operation of digital twins for infrastructure development);
- Computer vision (video analytics systems, infrastructure condition monitoring) [2].

According to the Autonomous Nonprofit Organization (ANO) "Digital Economy," AI was actively utilized across various sectors of urban life in Moscow in 2023. In terms of the number of active projects, transportation has 17 active projects, the economy has 16 active projects, digital technologies—15, utilities—12, construction, healthcare, and education—10 each, and safety—6 [7]. Examples of upcoming projects include autonomous taxis and trucks, the creation of a unified waste disposal system, a unified water resource management system, and a unified energy system.

The prospects and realities of AI as one of the key drivers of smart city development are undeniable. However, the rapid progress also reveals a layer of moral and ethical issues. Without addressing and resolving these, it is impossible to consider the creation of smart cities truly successful. It is necessary to develop ethical principles and standards of conduct when using AI systems [1]. The main priority is the protection of human rights (humanism and respect for autonomy and freedom, compliance with the law, non-discrimination, continuous risk assessment), responsibility for the consequences of creating, using, and applying AI (identification of AI when interacting with humans, data security, information security, control over AI self-improvement, risks of human oversight, and ensuring AI systems are used for their intended purposes), and transparency in AI communication (technology development beyond competition, competency development, collaboration among developers, and awareness of AI ethics).

References

1. Aleksakhin AV, Aleksakhina YuV, Gridchina AV, Dobryshina LN (2019) Innovative potential of intellectual organizations. Tver State University, Tver, Russia
2. Dobryshina LN, Gridchina AV (2018) State regulation of the economy in conditions of total digitalization. In: Terelyansky PV (ed) Step into the future: artificial intelligence and the digital economy. A revolution in management: a new digital economy or a new world of machines. Materials of the II international scientific forum. State University of Management, Moscow, Russia, pp 23–27
3. Gridchina AV, Aleksakhina YV (2021) Digital transformation of infrastructure facilities. In: Arkatova NA (ed) Vectors of the development of the modern economy: problems of theory and practice. Materials of the international scientific and practical conference. Moscow Polytechnic University, Moscow, pp 76–82
4. Gridchina AV, Aleksakhina YV (2022). Domestic experience of digitalization of the urban environment. In: Tumin VM, Senderov VL, Vatutina LA, Mazur VV, Lastovsky MA (eds) Current trends and prospects of managing socio-economic systems in the digital environment: materials of the international scientific and practical conference. Memory of the Honored Scientist of the Russian Federation V.I. Kravtsova. Moscow Polytechnic University, Moscow, Russia, pp 327–332

5. Kalinsky OI, Kruzhkova GV, Aleksakhin AV, Molchanov GA (2019) Selection of the optimal strategy for the supply of raw materials based on game theory. In Solovev D (ed) Smart technologies and innovations in design for control of technological processes and objects: economy and production. Springer, Cham, Switzerland, pp 577–583. https://doi.org/10.1007/978-3-030-18553-4_70
6. Official Portal of the Moscow Mayor and Moscow Government "Mos.ru" (2024) Merging of the real and the virtual. How does a smart city work? https://www.mos.ru/city/projects/smartcity/. Accessed 1 July 2024
7. Official Portal of the Moscow Mayor and Moscow Government "Mos.ru" (2023) The control system for the disposal of construction waste was recognized as the best at the Green Eurasia competition. https://www.mos.ru/news/item/124180073/. Accessed 5 July 2024
8. Presidium of the Government Commission on Digital Development, Use of Information Technologies for Improving Quality of Life and Conditions for Entrepreneurial Activity (2020) Passport of the federal project "artificial intelligence" (approved by Annex No. 3 to the minutes of August 27, 2020 No. 17). Moscow, Russia. https://sudact.ru/law/pasport-federalnogo-proekta-iskusstvennyi-intellekt-natsionalnoi-programmy/. Accessed 1 July 2024
9. Savon DY, Aleksakhin AV, Skryabin OO, Goodilin AA (2019) Occupational health and safety digitalization in the coal industry. Eurasian Min 2:70–72. https://doi.org/10.17580/em.2019.02.15
10. The Smart CityProject. (2024) About the project. https://russiasmartcity.ru/about. Accessed 3 July 2024
11. The Smart CityProject (2024) IQ of cities. https://russiasmartcity.ru/iq. Accessed 3 July 2024
12. Woetzel L, Remes J, Boland B, Lv K, Sinha S, Strube G, von der Tann V (2018) Smart cities: digital solutions for a more livable future [Report]. McKinsey. https://www.mckinsey.com/capabilities/operations/our-insights/smart-cities-digital-solutions-for-a-more-livable-future. Accessed 5 July 2024

Economic Characteristics and Principles of the Formation of the Transport Cluster in the Tourism Sector in the Conditions of the Digital Economy

Mamura N. Kushakova, Bekjan A. Akhmedov, Mukhlisa S. Kushakova, and Dilfuza R. Umarova

Abstract The research emphasizes the economic characteristics and principles of forming a transport cluster in the tourism sector in the digital economy and the importance of transport services in the further development of the tourism sector. Moreover, the research explores the possibilities of using a transport cluster in improving transport services in the tourism sector in the digital economy. The authors provide recommendations for more cost-effective information processing in the field of transport logistics. The authors carried out the analysis and provided recommendations to improve the quality of service and improve the economy of this area. A "Technical assignment" for the project was prepared and sent for approval to the Ministry of Information and Cyber Security of the Uzbek Republic. The system manages transport and logistics processes, using intelligent video surveillance to determine the specific number of rolling stock. This technology makes it possible to better identify freight wagons, automatically track their locations, and automatically record parking times on access roads while minimizing the need for human involvement.

Keywords Tourism · Transport · Cluster · Tourism and transport cluster · Transport infrastructure

JEL codes E650 · J150 · A120

M. N. Kushakova (✉) · D. R. Umarova
Tashkent State Transport University, Tashkent, Uzbekistan
e-mail: kushakova.m.tdtu@gmail.com

B. A. Akhmedov
Chirchik State Pedagogical University, Chirchik, Uzbekistan

M. S. Kushakova
Tashkent Pharmaceutical Institute, Tashkent, Uzbekistan

S. M. Sultanova et al. (eds.), *Sustainable Development of Transport*, Lecture Notes in Networks and Systems 1366, https://doi.org/10.1007/978-3-031-88846-5_13

1 Introduction

Thanks to the "Travel around Uzbekistan" program, the number of domestic tourists has increased to 12 million. Moreover, the number of foreign visitors has reached 9 million. These achievements align with the goals outlined in the New Uzbekistan strategy for sustainable development from 2022 to 2026 [1]. In light of this, the fast expansion of networks of roads and communication infrastructure is crucial to the realization of all goals pertaining to the modernization and renewal of the country, the establishment of an economically and qualitatively advanced Uzbek economy, and the provision of the all-encompassing development of regions in the country.

Tourism is widely regarded as one of the most powerful drivers of regional economic growth. Due to its positive impact on consumers, tourism and popular destinations play a key role in the success of municipal development plans. Many scholars, particularly those specializing in regional economies, argue that tourism can serve as a significant engine for development.

Because the traffic flow comprises many individual cars and is critical to achieving the region's tourism potential, transportation infrastructure is seen as a prerequisite for the growth of transportation services. Easy access to tourist destinations relies on the development of well-maintained and extensive transport routes, which, in turn, depends on the availability of a robust transport infrastructure.

Therefore, it is crucial to enhance the competitiveness of transportation businesses and organizations that serve tourists. As a key factor in ensuring economic efficiency, the significant potential of the cluster system, already tested in practice, must be carefully considered when addressing this issue. The main objective of this research is to explore the potential applications of the transport cluster within the tourism industry.

2 Methods

In today's world, the most competitive nations are mostly attributing their increased competitiveness to the cluster strategy [2, 3]. The creation of clusters as an organizational structure to ensure efficient collaboration among economic participants and as a tool for public–private partnerships is unfortunately under-discussed in Uzbekistan. However, it is worth noting that the term "cluster" is increasingly used by Uzbekistan's economic strategists when planning the future of various industries. By effectively utilizing clusters, Uzbekistan aims to address several pressing issues, including imbalances in the national transport system, the sector's low level of innovation, and a shortage of skilled workers.

One of the most essential components of the tourist industry, widely recognized as a driver of economic growth and development, is the cluster system for outsourcing and managing transport services [5, 6]. In this research, the authors aim to explore the processes that enhance regional competitiveness by applying the theoretical framework of establishing a transportation hub for the tourism sector. Additionally, the authors provide practical recommendations to support these findings [4].

3 Results and Discussion

In recent years, many scholars have studied clusters based on the specific characteristics of different regions. Clusters impact business activities, regional growth, and national competitiveness, making them a popular area of research. As noted by researcher Michael Porter, clusters are closely tied to the concept of competitiveness because they drive innovation, enhance customer service, foster collaboration with other institutions, and coordinate the demand for information and knowledge to develop new technologies. Porter defines a cluster as a geographically proximate group of interconnected businesses and organizations within the same industry that share resources and collaborate effectively.

According to Bergman and Feser, clusters are groups of businesses, organizations, or even unrelated companies, where membership in the group enhances a company's competitiveness [2].

Martin and Sunley argue that a precise definition of a cluster is impossible due to the complexity and fuzziness of the concept. When looking at the big picture, clusters may also be described as groupings of businesses in the same industry that are located in the same broad region [5].

The goal of a cluster in the tourism industry is to bring together enterprises that typically operate independently to create a successful tourism product for the region. As tourism increasingly becomes a driver of economic growth, especially at the regional level, the tourism cluster is composed of key services such as lodging, food and beverage preparation and service, passenger transportation, tour guides and other tourism professionals, property rentals, consultancy, specialized education and training, and other related commercial activities (Table 1).

The concept of cluster economics has gained popularity in developed and developing nations, driven by the increasing globalization of economic processes. As a result, various clusters have emerged across the world. Table 1 highlights examples from the USA, Europe, and Asia, showcasing how different countries have utilized clusters to enhance their economies.

As an organizational approach to supply chain management, clustering comes with its own set of risks, benefits, and challenges. For more details, refer to Fig. 1.

This research also explores the tourism potential of the Khorezm Region, considering visitors' needs, even though many local and international scholars have already examined various aspects of cluster theory and the development of transport-logistics systems.

Samuylov and Pokrovskaya define a "transport-logistics cluster" (TLK) as a group of related businesses in a specific region that collaborate by sharing resources, a central hub, and a network of service providers to achieve common goals. Cluster

Table 1 American, European, and Asian models of cluster use

The main directions	Fields of application of clusters
State support	The growth of scientific and technological collaboration, the provision of favorable tax policies, capital raising, and coordinated efforts of state institutions, industrial businesses, and scientific groups
Business model	Maximum use of natural, scientific, and human resources, as well as the concentration of businesses in one place
Cluster types	Industry 4.0, motion picture, automobile, and space sectors
European model of cluster development	
State support	Reducing taxes on investments in R&D, designing regulations specifically aimed at fostering innovative activities, allocating budget funds to support high-tech businesses, and encouraging R&D initiatives that address industrial needs
Business model	Business aggregation around a single, dominant firm
Cluster types	Perfume, cosmetics, textiles, food, wine, etc.
Germany	
State support	Integrating research facilities with businesses. Clusters may be classified into three types: scientific, which leverages its technology in production; clusters that integrate research and production; and clusters that are based on innovative enterprises. Federal and municipal sources provide financial resources via industrial clusters
Business model	Small, medium, and big businesses interacting
Cluster types	Chemical, printing, and transport equipment manufacturing cluster, automotive industry cluster
The Asian model of cluster development (in the case of Japan)	
State support	Encourage the development of highly specialized areas and attraction of advanced technologies from abroad
Business model	The Japanese cluster is a two- or three-level subcluster that is geographically close to it
Cluster types	Electrical engineering, automobile engineering, etc.

Source Developed by the authors

activities aim to leverage the core strengths of members in the area to maximize the synergistic effects of coordinated efforts.

To maximize the synergistic effects of well-coordinated efforts, cluster activities aim to draw on the primary strengths of cluster members in the area. Recent scientific literature described a transport cluster as an interdependent network of organizations involved in the supply, production, consumption, and research of goods and services related to industrial infrastructure.

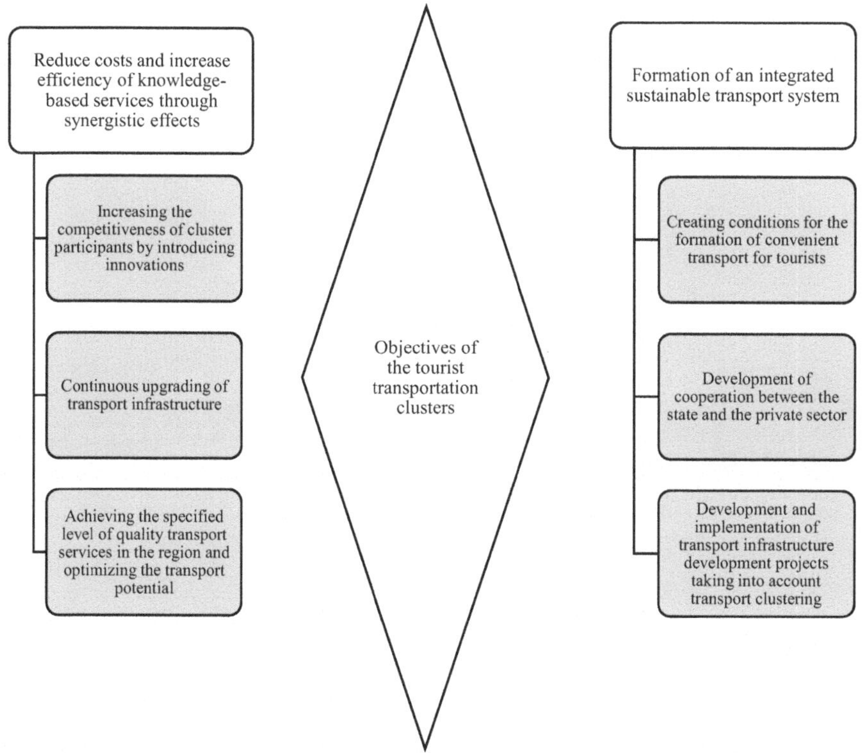

Fig. 1 Objectives of the tourist transportation cluster. *Source* Developed by the authors

Developing a supportive infrastructure and facilitating cluster development procedures are the primary objectives of regional authorities as they carry out the regional cluster development plan. Creating transportation and logistics clusters that pave the way for the region's tourism industry to thrive can help achieve a lot (Fig. 2).

The examination of the cluster properties showed that the main priorities of the transportation logistics group should be as follows:

- Improvement and maintenance of transportation networks, including highways, railways, waterways, seaports, parking facilities, and other roadside amenities;
- Establishment of an advanced system of multimodal terminals and logistics hubs;
- Enhancement of the processes for constructing, repairing, and maintaining highways, with a focus on the efficient use of these roads.
- Geographic expansion of transportation infrastructure;
- Ensuring uniform technological compatibility across major transportation routes and fostering interconnection between different types of transport;
- Promoting innovation and the introduction of new technologies and intelligent systems in the transport and logistics industry.

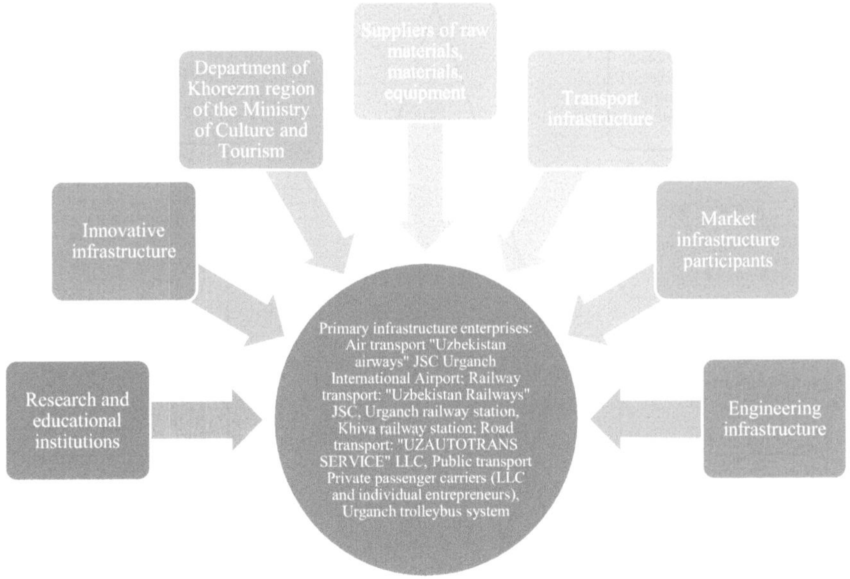

Fig. 2 Model of tourist transport cluster organization in Khorezm region. *Source* Developed by the authors

In summary, tourist transportation activities can be more effectively organized using a cluster approach. This approach involves restructuring enterprise management within the current system around an integrative factor, aligning the core activities with surrounding infrastructure systems to work toward a common goal.

From this perspective, this research proposes a cluster approach to organizing the transport system in the Khorezm Region. The primary objective is to establish a well-coordinated transportation network that serves domestic and international tourists. The transport cluster would include enterprises involved in passenger transport, Urganch Central Railway Station, Urganch International Airport, and other related services, creating a unified system to support tourism in the region.

In addition to the main participants already mentioned, the initial structure of the cluster includes transport enterprises, tourism businesses, educational institutions, research institutes, engineering infrastructure, financial organizations, and information and innovation infrastructure. These entities play a vital role in the efficient organization of domestic and international tourist transportation activities. The businesses within the cluster are responsible for designing and providing transportation services to passengers, ensuring seamless travel experiences for tourists.

Thus, the structures affecting the activity of transport enterprises, which are part of the primary structure of the transport cluster of the Khorezm Region, include the following:

- Transport infrastructure (motorways, railways, airports, and transport enterprises);

- Market infrastructure participants (banks and leasing and marketing companies);
- Engineering infrastructure (telecommunications and data transmission, vehicle service providers, repair shops, energy, gas, water, heat supply and waste disposal);
- Innovative infrastructure organizations (consulting services, research centers, venture capital firms or funds, business incubators, innovation centers, and IT parks);
- Research and educational institutions (secondary education, vocational education, higher education institutions, and industrial and research institutes).

The model for selecting the optimal structure and composition of regional cluster participants is based on the principle of recognizing the cluster's properties as a uniquely designed framework. It emphasizes the distinctive characteristics of the cluster-model approach to integrating participants, ensuring that the specific needs and dynamics of the cluster are effectively addressed.

Connecting the online ticketing system of JSC "Uzbekistan Railways" to the single electronic site UzTrans.uz is an improvement to the current system that makes it possible to sell electronic tickets for air, railway, and intercity and interregional buses. The developers of the unified electronic platform of a transport system were provided with the necessary technical specifications for integration (API).

As part of the development of the "Passenger Service Center" digital platform, an information system for dispatching and monitoring high-speed trains has been created. The technical terms of reference for this project were prepared and submitted to the Ministry of Information Technologies and Communications Development on September 9, 2021 (Ref. 02/880-21). Subsequently, on February 8, 2019, these specifications were forwarded to the Cyber Security Center for approval (Ref. 21).

4 Conclusion

A technical assignment for the project was prepared and sent for approval to the Ministry of Information and Cyber Security of the Uzbek Republic. The system manages transport and logistics processes, using intelligent video surveillance to determine the specific number of rolling stock. With this technology, freight wagons can be better identified, their whereabouts can be automatically monitored, and the time it takes to park on access roads can be recorded automatically while reducing the need for human intervention.

Current operations at the Sergeli and Nazarbek stations are in test mode. This information system has also been fully integrated with the software of the electronic railway scales that were installed at the Keles station.

Many of the steps planned for implementation this year are currently underway.

A substantial investment is required to digitize the transportation system, and leading nations are expected to continue investing heavily in transportation digitalization in the foreseeable future. In this context, Russia has laid the groundwork for digitization through the development of several interoperable platforms. Developed countries, which were early adopters of the Internet, are now at the forefront of integrating digital technology into their transportation systems and other sectors of their economies. The initial advancements in transportation digitalization were made in the United States and the United Kingdom, with subsequent adaptations in countries such as China, South Korea, and India. It would be beneficial to apply the experiences of these nations to Uzbekistan's context, as monitoring their progress can facilitate the transfer of valuable lessons on transportation digitalization.

In addition to using ICTs, a complete digitization of the economy, including transportation, requires a complete reorganization of the country's economic infrastructure.

Participants in the cluster model should focus on providing a stable and well-organized system, along with innovative solutions to problem-solving, particularly in replicating new information, technologies, and products. This approach fosters a system of interconnected, research-based technology networks. Internal specialization and the ability to reduce innovation and standardization costs give cluster participants additional competitive advantages. Examples of such businesses can vary widely, from specialized service clusters targeting specific markets to small enterprises driving the growth of new economic sectors within the local economy.

References

1. President of the Republic of Uzbekistan (2022) Decree "On the development strategy of the new Uzbekistan for 2022–2026" (January 28, 2022 No. DP-60), Tashkent, Uzbekistan. https://lex.uz/docs/6968143. Accessed 17 Sept 2024
2. Bergman EM, Feser EJ (1999) Industry clusters: a methodology and framework for regional development policy in the United States. In: Boosting innovation: the cluster approach. OECD, Paris, France, pp 243–268. https://research.manchester.ac.uk/en/publications/industry-clusters-a-methodology-and-framework-for-regional-develo. Accessed 17 Sept 2024
3. Khodzhiniyazov ES (2023) Characteristics of the formation of transport clusters in the tourism sector. In: Mamun academy of Khorezm, Khiva, Uzbekistan, pp 1–3
4. Kovačić M, Milošević T (2016) Interdependence of transport and tourism. Pomorski Zbornik 52(2016):99–111. https://hrcak.srce.hr/file/251481?ref=f-zin.faktograf.hr. Accessed 17 Sept 2024
5. Martin R, Sunley P (2003) Deconstructing clusters: chaotic concept or policy panacea? J Econ Geogr 3(1):5–35. https://www.jbs.cam.ac.uk/wp-content/uploads/2023/05/cbrwp244.pdf. Accessed 17 Sept 2024
6. Porter ME (1998) Clusters and competition: new agendas for companies, governments, and institutions. In: Porter ME (ed) On competition. Harvard Business School Press, Boston, MA, pp 197–299

The Role of Globalization and Intercultural Communication in the Digitalization of the Economy and Transportation

Salaxiddin I. Ismailov, Guldora S. Mustaeva, Gulorom T. Tulaboeva, and Dilfuza Z. Saydaliyeva

Abstract The research aims to define the role of globalization and intercultural communication in the digitalization of the economy and transportation. Utilizing correlation and regression analysis, the authors conduct economic-mathematical modeling to determine the influence of identified socio-cultural factors on the digitalization of the economy and transportation, drawing upon advanced international experience from 27 countries. The official statistical data from IMD for 2023 serves as the information-empirical basis. As a result, the authors formulated economic-mathematical regularities of digital development of the economy and transportation under the influence of socio-cultural factors. Additionally, the authors developed a new model of the digital transport economy and provided recommendations for its implementation in the Republic of Uzbekistan. Moreover, the authors determined the perspective of digitalization of the economy and transportation through optimizing the influence of socio-cultural factors on them in the long term of the Fourth Industrial Revolution in the new model of the digital transport economy. The main conclusion is that globalization and intercultural communication play a significant role in the digitalization of the economy and transportation. The theoretical significance of the obtained results lies in refining the composition of key socio-cultural factors and quantitatively measuring their influence on the digitalization of the economy and transportation. The practical significance of the developed model of the digital transport economy is expressed in its ability to improve the contemporary practice of state regulation of international trade development. The managerial significance of the author's recommendations lies in their potential to strengthen the foreign trade cooperation of the Republic of Uzbekistan with the Eurasian Economic Union.

S. I. Ismailov (✉)
Samarkand State Institute of Foreign Languages, Samarkand, Uzbekistan
e-mail: salakhiddinismailov49@gmail.com

G. S. Mustaeva · G. T. Tulaboeva · D. Z. Saydaliyeva
Tashkent State Transport University, Tashkent, Uzbekistan

S. M. Sultanova et al. (eds.), *Sustainable Development of Transport*, Lecture Notes in Networks and Systems 1366, https://doi.org/10.1007/978-3-031-88846-5_14

Keywords Globalization · Cultural openness · Intercultural communication · Digitalization of the economy and transportation · Digital transport economy · International trade

JEL codes F01 · L96 · L98 · R48 · Z18

1 Introduction

The digitalization of the economy and transportation opens up new opportunities for developing international trade. First, the application of digital technologies expands the boundaries of exports. Through e-commerce, manufacturers can remotely accept orders from around the world for their products and transport them at low costs. This makes exports more accessible, profitable, and geographically diversified, providing economies of scale.

Second, the digital transition of the transport and logistics complex takes the optimization of transnational supply chains to an even higher level. Importing raw materials and materials, components, and certain types of finished products becomes more advantageous than ever. This enables a more complete realization of the potential of international division of labor. The combination of various modes of transport becomes even more accessible and straightforward. The utilization of the production capacities of transport and logistics companies becomes more complete, enhancing their operational efficiency.

Third, digital modernization of the transport and logistics complex makes it possible to create trade routes of the Fourth Industrial Revolution. Digital transport is safer, more predictable, and manageable. The expanding array of digital transport and logistics services enhances the attractiveness of trade routes, speeding up delivery times and reducing the risks of transporting goods. Undoubtedly, digital transport is preferable for transporting products of high-tech industries because they are integrated into the production and distribution system of Industry 4.0.

Although the importance of the digital development of the economy and transportation is widely recognized, the causal relationships of this process remain largely uncertain. Financial (government funding and private investments) and technological factors (the progressiveness and accessibility of digital technologies for the transport and logistics complex) are the most thoroughly studied and widely discussed factors of the digitalization of the economy and transportation. However, these factors do not fully explain or determine the digital development process of the economy and transportation.

Meanwhile, socio-cultural factors remain under-researched and unclear, although their role in the digitalization of the economy and transportation is potentially significant. The proposed assumption is based on the premise that socio-cultural factors determine the staffing and social support of the examined process. This research aims to fill the identified gap in the literature and seeks to determine the role of globalization and intercultural communication in the digitalization of the economy and transportation.

2 Literature Review

This research builds upon the theory of the digitalization of the economy and transportation, which has laid a fundamental basis for forming a contemporary model of the digital transport economy [1, 11]. In the existing model, the key role in the digitalization of the economy and transportation is attributed to financial and technological factors [3, 7, 8]. According to the established model, the following socio-cultural factors of digital development of the economy and transportation are identified:

- Digital skills, particularly digital personnel [6];
- Cultural openness of society to globalization [2];
- Language skills of intercultural communication [10];
- Level of proficiency in the English language in society [12].

However, the mentioned theory does not explain the scale and nature of the influence of the listed socio-cultural factors on the digitalization of the economy and transportation, which constitutes a gap in the literature. Due to this gap in the contemporary model of the digital transport economy, socio-cultural factors are assigned a secondary role. This diminishes the manageability of the digital transport economy and hampers its development, representing a significant drawback of its current model. Therefore, filling the identified gap in the literature is necessary, which has determined the formulation of the following research question (RQ): What impact do socio-cultural factors have on the digitalization of the economy and transportation?

Based on the evidence provided in the literature by He and Lee [4] and Tiran et al. [9] that socio-cultural factors play an important role in the innovative development of the economy (through social support or social resistance to changes), the following hypothesis (H) is proposed: socio-cultural factors largely determine the digitalization of the economy and transportation.

To test hypothesis H, the authors conduct economic-mathematical modeling of the influence of identified socio-cultural factors on the digitalization of the economy and transportation, drawing upon advanced international experience.

3 Materials and Methods

To determine the role of globalization and intercultural communication in the digitalization of the economy and transportation, this research employs correlation and regression analyses. The informational and empirical basis of this research consists of data from the official statistics of IMD for 2023 [5]. The sample includes 27 countries for which IMD provided all necessary and openly accessible statistics.

The research is conducted in three consecutive stages. In the first stage, the relationship (correlation) and the influence of the factor "Development and application of tech." (DAT) on the development of the transport and logistics complex are determined, particularly in road transport ("Roads," RDS), railway transport ("Railroads," RLR), and air transport ("Quality of air transportation," QAT). In the second stage, the relationship (correlation) and dependencies of DAT on socio-cultural factors are identified, including the following:

- "Digital/Technological skills" (Fsc_1);
- "Attitudes toward globalization" (Fsc_2);
- "Language skills" (Fsc_3);
- "English proficiency—TOEFL" (Fsc_4).

The statistics for this research are summarized in Table 1.

In the third stage, the prospect of digitalizing the economy and transportation is determined by optimizing the influence of socio-cultural factors on them in the long term of the Fourth Industrial Revolution. For this purpose, the obtained regression models are supplemented with the maximum possible values of selected socio-cultural factors that positively influence DAT. By means of trend analysis, the changes (growth) in the values of all variables are identified following the optimization compared to the current indicator values in 2023.

4 Results

To quantitatively measure the impact of globalization and intercultural communication on the digitalization of the economy and transportation, the authors conducted correlation and regression analysis of statistical data from Table 1. In the first stage of this research, the authors established a fairly close relationship between the DAT factor and the development of the transport and logistics complex, particularly in road transport (correlation: 30.35%), railway transport (correlation: 10.69%), and air transport (correlation: 58.95%). The regression dependencies were as follows:

$$RDS = -3.31 + 0.76 * DAT \tag{1}$$

$$RLR = -0.005 + 0.007 * DAT \tag{2}$$

$$QAT = 0.86 + 0.94 * DAT \tag{3}$$

The regression Eqs. (1), (2) and (3) indicate that with an increase in the activity of applying digital technologies (DAT) by 1 point, the density of the road transport network increases by 0.76 km of roads per square km of land area; the density of the railway transport network increases by 0.007 km per square km; and the number of passengers carried by main airline companies increases by 0.94 thousand.

Table 1 Digitalization of the economy and transport and factors of globalization and intercultural communication sample countries in 2023

Country	Digital/techno-logical skills, scores 1–10	Attitudes toward globalization, scores 1–10	Language skills, scores 1–10	English proficiency—TOEFL, scores 1–100	Roads (density of the network), km roads/square km land area	railroads (density of the network), km per square km	Quality of air transportation (number of passengers carried by main companies), thousands	Development and application of tech., scores 1–10
Bahrain	8.46	7.23	8.52	92.00	5.55	0.00	7.84	7.58
Bulgaria	6.31	4.47	5.40	95.00	0.18	0.04	5.77	5.37
China	7.74	7.73	6.39	87.00	0.56	0.01	8.44	6.98
Croatia	7.00	5.03	6.91	96.00	0.47	0.03	6.21	4.62
Czech Republic	7.26	7.28	7.15	95.00	1.66	0.12	7.87	6.60
Estonia	6.60	6.84	7.68	96.00	1.30	0.03	4.68	7.30
Finland	8.50	8.03	8.36	96.00	0.23	0.02	8.48	8.28
France	6.94	3.87	4.86	89.00	2.03	0.05	8.15	6.25
Iceland	8.12	7.67	8.63	97.00	0.13	0.00	8.33	7.67
Indonesia	6.89	7.25	6.31	88.00	0.29	0.00	7.01	7.13
Ireland	7.06	8.20	6.16	94.00	1.44	0.03	7.53	7.22
Japan	4.65	5.82	3.51	73.00	3.39	0.05	8.58	5.91
Kazakhstan	5.78	6.38	5.36	81.00	0.06	0.01	5.90	6.54
Korea Rep	6.18	7.82	6.11	86.00	1.13	0.04	8.42	5.76
Lithuania	8.69	6.44	7.68	93.00	1.30	0.03	5.52	6.75
Malaysia	7.12	6.62	6.68	92.00	1.76	0.01	7.61	6.81
Mongolia	6.00	5.75	4.67	84.00	0.07	0.00	3.46	3.89

(continued)

Table 1 (continued)

Country	Digital/ techno-logical skills, scores 1–10	Attitudes toward globalization, scores 1–10	Language skills, scores 1–10	English proficiency —TOEFL, scores 1–100	Roads (density of the network), km roads/ square km land area	railroads (density of the network), km per square km	Quality of air transportation (number of passengers carried by main companies), thousands	Development and application of tech., scores 1–10
Philippines	6.21	6.89	7.10	90.00	0.72	0.00	5.47	5.64
Poland	6.35	5.20	5.44	92.00	1.38	0.06	6.68	5.85
Portugal	7.26	6.79	8.18	94.00	0.16	0.03	6.59	6.54
Qatar	8.14	7.14	7.91	82.00	1.52	0.02	8.84	7.94
Singapore	7.76	7.56	7.80	98.00	13.03	0.34	8.41	7.66
Slovak Republic	7.18	5.03	5.54	95.00	0.37	0.07	5.23	5.38
Slovenia	7.00	5.50	7.80	100.00	1.92	0.06	2.99	5.96
Thailand	6.74	7.65	5.38	83.00	1.37	0.01	7.84	6.73
Turkey	6.20	6.75	5.14	88.00	0.33	0.01	7.10	5.88
UAE	7.48	8.11	8.38	85.00	1.18	0.00	9.10	7.34

Source Compiled by the authors based on materials from IMD [5]

In the second stage of this research, a strong correlation between DAT and the socio-cultural environment was identified (multiple correlation: 79.17%). The regression dependency was as follows:

$$DAT = 2.44 + 0.48 * Fsc_1 + 0.36 * Fsc_2 + 0.12 * Fsc_3 - 0.03 * Fsc_4 \quad (4)$$

The regression Eq. (4) signifies that with an increase in the level of development of digital skills in society by 1 point, the activity of applying digital technologies (DAT) increases by 0.48 points. With an increase in the level of cultural openness of society to globalization by 1 point, the activity of applying digital technologies (DAT) increases by 0.36 points.

With an increase in the level of proficiency in language skills of intercultural communication by 1 point, the activity of applying digital technologies (DAT) increases by 0.12 points. With an increase in the level of proficiency in the English language in society by 1 point, the activity of applying digital technologies (DAT) decreases by 0.03 points. Therefore, mastering the English language hampers the development and dissemination of digital technologies and the digitalization of the economy and transportation.

In the third stage of this research, a new model of the digital transport economy has been developed based on the results of economic-mathematical modeling. The novelty of the new model lies in its focus on managing socio-cultural factors and transitioning from mastering the English language to mastering the national languages of the countries' trading partners.

The prospect of digitalization of the economy and transportation has been determined by optimizing the influence of socio-cultural factors on them in the long term of the Fourth Industrial Revolution in the new model. For this purpose, the maximum possible values of selected socio-cultural factors ($Fsc_1 - Fsc_3$, equal to 10 points) positively influencing DAT were inserted into the obtained regression models. By means of trend analysis, the changes (growth) in the values of all variables following the optimization were determined compared to the current indicator values in 2023 (Fig. 1).

According to Fig. 1, with the transition to the new model, the perspective of digitalizing the economy and transportation is associated with a 46.83% increase in the activity of applying digital technologies (DAT) (from 6.50 points in 2023 to 9.55 points). This will provide the following advantages in the development of the transport and logistics complex:

- The increase in density of the road transport network rises by 143.09% (from 1.61 to 3.92 km of roads per square km of land area);
- The increase in density of the railway transport network rises by 52.90% (from 0.04 to 0.06 km per square km);
- The growth in the number of passengers carried by main airline companies increases by 41.047% (from 6.96 to 9.82 thousand).

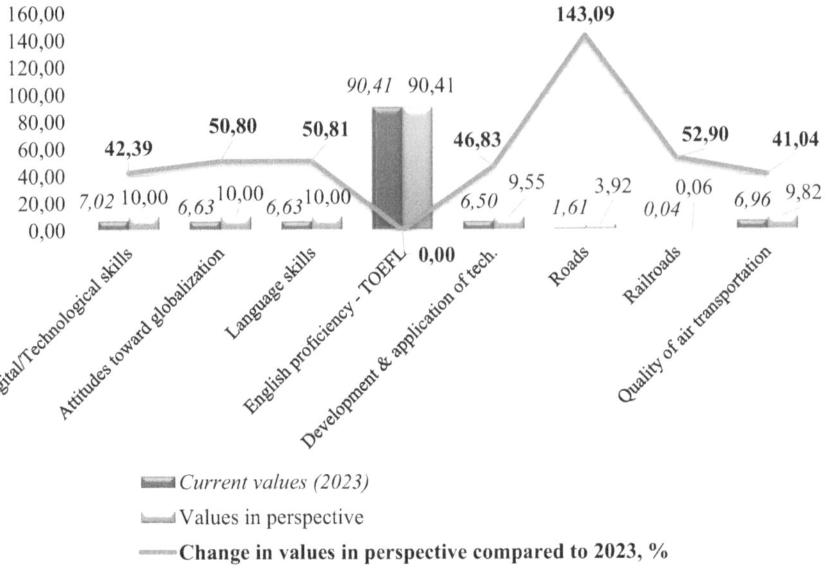

Fig. 1 Perspective of digitalization of economy and transport by optimizing the influence of socio-cultural factors on them. *Source* Calculated and compiled by the authors

To optimize the influence of socio-cultural factors on the digitalization of the economy and transportation, a complex of the following quantitative and qualitative recommendations is proposed:

- Increase the level of cultural openness of society to globalization by 42.39% %;
- Enhance the level of development of language skills of intercultural communication by 50.80%;
- Raise the level of proficiency in the English language in society by 50.81%;
- Redirect from studying the English language to studying the national languages of foreign trade partners. For example, to develop foreign trade cooperation between the Republic of Uzbekistan and the Eurasian Economic Union (EAEU), it is suggested to introduce the study of the Armenian, Belarusian, Kazakh, Kyrgyz, and Russian languages in Uzbekistan.

5 Discussion

The contribution of the research to the literature [1, 11] lies in developing the theory of digitalization of the economy and transportation by refining the role of globalization and intercultural communication in the digitalization of the economy and transportation. The following significant and positive contributions to the digital development of the economy and transportation have been demonstrated:

- The development of digital skills, particularly the training of digital personnel (in support of Serova et al. [6]);
- The increase in cultural openness of society to globalization (in support of Bogoviz et al. [2]),
- The development of language skills for intercultural communication (in support of Tovmasyan et al. [10]).

Considering the identified important role of socio-cultural factors (hypothesis H is confirmed in support of He and Lee [4] and Tiran et al. [9], a new model of digital transportation economy has been developed. In contrast to the existing model, the new model is characterized by the following:

- The key role in the digitalization of the economy and transportation is attributed not to financial and technological factors (in contrast to Galoyan and Matevosyan [3], Sultanova [7], and Sultanova and Babakhanova [8]) but to socio-cultural factors;
- Due to the demonstrated negative impact of mastering the English language on the digitalization of the economy and transportation (in contrast to Wang [12]), the new model suggests the acquisition of the national languages of countries that are foreign trade partners instead.

6 Conclusion

Thus, the main conclusion of the conducted research is that globalization and intercultural communication play a significant role in the digitalization of the economy and transportation (hypothesis H is confirmed). The following results have been obtained from the research:

- Economic and mathematical regularities of digital development of the economy and transportation under the influence of socio-cultural factors were identified;
- A new model of digital transportation economy was developed; recommendations for its implementation in the Republic of Uzbekistan were proposed;
- The perspective of digitalizing the economy and transportation through optimizing the influence of socio-cultural factors in the long term of the Fourth Industrial Revolution in the new model of digital transportation economy was determined.

The theoretical significance of the obtained results lies in refining the composition of key socio-cultural factors and quantitatively measuring their influence on the digitalization of the economy and transportation. The practical significance of the developed model of the digital transportation economy is expressed in its ability to improve the current practice of state regulation of international trade development. The managerial significance of the author's recommendations lies in their ability to strengthen the foreign trade cooperation of the Republic of Uzbekistan with the Eurasian Economic Union (EAEU).

References

1. Audonin A, Turginbayeva A, Askerov A, Yergobek D (2020) Modern economic and logistical trends in Eurasia: how do new trans-Eurasian mega-projects influent to national economic growth. In: E3S web of conferences, vol 159, p 06009. https://doi.org/10.1051/e3sconf/202015906009
2. Bogoviz AV, Shvakov EE, Tretyakova OG, Zakharov MY, Abramov AN (2020) Globalization of education in the conditions of formation of the global knowledge economy: regularities and tendencies. In: Popkova, E (ed) Growth poles of the global economy: emergence, changes and future perspectives. Springer, Cham, Switzerland, pp 993–1000. https://doi.org/10.1007/978-3-030-15160-7_99
3. Galoyan D, Matevosyan D (2023) The impact of the Russian-Ukrainian conflict on Armenia's foreign trade: a quantitative analysis. J Lib Int Aff 9(2):89–103. https://doi.org/10.47305/JLIA2392106g
4. He M, Lee J (2020) Social culture and innovation diffusion: a theoretically founded agent-based model. J Evol Econ 30(4):1109–1149. https://doi.org/10.1007/s00191-020-00665-9
5. IMD (2024) World competitiveness online. https://worldcompetitiveness.imd.org/. Accessed 7 March 2024
6. Serova OA, Golikov VV, Petrenko ES, Stolyarov NO (2020) Communication models of AI as manifestations of its social origin. In: Popkova E, Sergi B (eds) Artificial intelligence: anthropogenic nature vs. social origin. Springer, Cham, Switzerland, pp 331–337. https://doi.org/10.1007/978-3-030-39319-9_37
7. Sultanova SM (2021) Improvement of financial flow management of a railway transport enterprise. Turk Online J Qual Inq 12(6):7117–7121
8. Sultanova SM, Babakhanova NU (2023) Statistical model for determining the quality of cargo work JSC "Uzbekistan Railways." In: E3S web of conferences, vol 402, p 06016. https://doi.org/10.1051/e3sconf/202340206016
9. Tiran J, Bole D, Kozina J (2023) Industrial culture as an agent of social innovation: reflections from Velenje, Slovenia. Innov: Eur J Soc Sci Res 36(2):333–356. https://doi.org/10.1080/13511610.2022.2026212
10. Tovmasyan G, Hakhverdyan D, Mkrtchyan T, Petrosyan A (2022) The impact of COVID-19 on education quality and learning efficiency in universities: evidence from Armenia. Issues Educ Res 32(4):1623–2164. http://www.iier.org.au/iier32/tovmasyan.pdf. Accessed 12 March 2024
11. Tursunov B (2022) Financial security in small business in period of digital economy: in case of Uzbekistan. In: Proceedings of ICFNDS '22: the 6th international conference on future networks and distributed systems. Association for Computing Machinery, New York, NY, pp 491–498. https://doi.org/10.1145/3584202.3584275
12. Wang X (2022) International trade English phrases and grammar translation. Sci Program 2022:3726492. https://doi.org/10.1155/2022/3726492

AI Integration in Smart Car Parking Stations

Gulira'no Kh. Khalilova, Malika Sh. Choriyeva, Rustam G. Samatov, Umidulla A. Abdurazzokov, and Dilshod A. Urunov

Abstract The research delves into the pressing issue of parking shortage in urban areas and explores solutions by integrating artificial intelligence (AI) in smart car parking stations. The authors begin by analyzing the results of a survey of mobile application users to understand their parking habits and preferences. A survey was conducted to study the opinions of mobile application users for parking. The survey asked drivers about the time they spend searching for a parking space, how often they use parking spaces in city centers, and whether they would like to use a mobile app. Subsequently, the authors discuss the theoretical design, advantages, and disadvantages of a parking system based on a mobile application. The integration of AI (e.g., predictive analytics, computer vision, and machine learning) is highlighted for its potential to optimize parking management and enhance urban mobility. The research underscores the transformative impact of AI-driven solutions on addressing congestion, pollution, and inefficiencies in parking operations.

Keywords AI · Smart car parking · Urban mobility · Congestion · Sustainability · Machine learning · Predictive analytics · Computer vision · Parking management mobile application · Parking system · Smart city · Telematics

JEL codes R4 · R41 · R42

1 Introduction

In the era of urbanization, the proliferation of vehicles has exacerbated the challenge of parking space management, leading to congestion, pollution, and frustration among commuters. Smart car parking stations emerge as a promising solution, integrating advanced technologies to optimize parking operations and enhance user experience. This research explores the integration of artificial intelligence (AI) in smart car parking stations, delineating its transformative impact on urban mobility.

G. Kh. Khalilova (✉) · M. Sh. Choriyeva · R. G. Samatov · U. A. Abdurazzokov · D. A. Urunov
Tashkent State Transport University, Tashkent, Uzbekistan
e-mail: xalilovagulirano444@gmail.com

S. M. Sultanova et al. (eds.), *Sustainable Development of Transport*, Lecture Notes in Networks and Systems 1366, https://doi.org/10.1007/978-3-031-88846-5_15

Through comprehensive research and analysis, the authors elucidate the principles, benefits, and challenges associated with AI-enabled parking solutions. Moreover, they highlight innovative AI applications, such as predictive analytics, computer vision, and machine learning algorithms, driving efficiency, sustainability, and convenience in parking management. The research also discusses the implications of AI integration in mitigating traffic congestion, reducing carbon emissions, and fostering sustainable urban development. By examining case studies and industry advancements, it provides valuable insights into the future trajectory of AI-driven smart car parking stations and their role in shaping the cities of tomorrow. Urbanization is accelerating at an unprecedented pace, with more than half of the world's population residing in cities. Alongside this rapid urban growth, the proliferation of vehicles has intensified the strain on transportation infrastructure, particularly in terms of parking space management. In congested urban areas, finding parking spots often becomes a daunting task, leading to traffic gridlock, environmental pollution, and wasted time and resources. In response to these challenges, smart car parking stations have emerged as a transformative solution, leveraging advanced technologies to optimize parking operations and improve the overall mobility experience. Traffic jams, wasted time of drivers looking for a place, emission of harmful gases into the atmosphere, and similar problems are occurring in the central cities of developed and developing countries due to the lack of parking spaces or lack of amenities. Solutions to these problems are presented in many articles and theses [1–3]. Nowadays, it is becoming difficult to find free parking spaces in our cities. Scientists have shown ways to further improve and increase the efficiency of parking lots through advanced technology and equipment. For example, it is possible to receive, analyze, and control data from a parking lot with sensors and cameras. These technological devices are said to partially solve the above problems [4, 5].

During this research, the authors came across many studies. Most of these studies are about the fact that the number of vehicles is increasing annually and, as a result, road traffic accidents occur that end in the tragic death of the population or various injuries as a result of traffic jams. There is extensive coverage of such unfortunate situations as the release of toxic gases from vehicles, which are considered harmful to the environment. The materials that inspired the writing of this research include "Machine learning based efficient and secured car parking system" by Krishnan et al. [6].

The introduction of the research provides a comprehensive overview of the growing significance of car usage in today's society, particularly in India. It effectively sets the stage by highlighting the increase in car ownership, the challenges posed by insufficient parking spaces, and the rising concerns of vehicle theft. By framing the discussion around these pressing issues, the research emphasizes the need for innovative solutions to address them.

The proposed mobile application-based parking system is introduced as a novel approach to tackling the challenges of parking space shortage and vehicle theft. The system's features include advanced booking, payment options, and security measures. These features are outlined in detail and demonstrate a thorough understanding of the problem and the potential solutions.

The related works section provides valuable context by reviewing existing research in intelligent parking systems. By citing various studies, the research showcases the evolution of parking technology and highlights the diverse approaches taken by researchers to address similar challenges.

The detailed explanation of the proposed system's workflow, from parking slot booking to exit procedures, helps readers grasp the functionality and potential benefits of the system. Integrating technologies such as Radio Frequency Identification (RFID), QR codes, and sensors underscores the system's sophistication and effectiveness in ensuring security and efficiency.

The section "Results and Discussion" presents empirical evidence supporting the effectiveness of the proposed system, including increased parking efficiency and reduced average parking time. The inclusion of data visualizations and analytical results adds credibility to the findings and reinforces the article's central arguments.

Overall, the research presents a well-structured and compelling case for adopting a mobile application-based parking system. It effectively addresses the challenges posed by insufficient parking space and vehicle theft while offering practical solutions and empirical evidence to support its claims.

More than 75% of waste production, 80% of emissions, and 75% of electricity use are accounted for by our cities today. About 20% of the total CO_2 emission in Europe is caused by vehicle emissions; the remaining 40% is urban mobility. The data shows that 30% of daily traffic congestion in highly urbanized areas occurs during the search for an empty place for a car [7]. Accordingly, reducing traffic congestion in cities can be achieved by developing intelligent parking equipment, telematics tools, and programs that help reduce the time spent searching for a place. Intelligent smartphone applications used in the field of parking spaces are one of the main parts of the "City" project [8, 9].

Automobile settlements made up most of the city's territory. For example, 81% of Los Angeles is provided with parking spaces. This indicator is slightly lower in European cities [10]: 23% in Paris, 23% in Munich, 19% in Copenhagen, and 18% in Hamburg are provided with parking spaces in relation to the total area. Therefore, we must apply scientific and practical methods to use parking lots efficiently.

Using the "Smart Parking" system will reduce traffic jams on the streets. If drivers find the accommodation as soon as possible, they will not spend too much time and fuel searching. This will prevent environmental damage.

Parking lot problems cannot be solved without advanced technologies. The research presents solutions that help solve parking problems. The opinion of the users of the parking lot in Tashkent is also presented.

At the forefront of this technological evolution is the integration of AI, which empowers smart parking facilities with intelligent capabilities to streamline operations, enhance efficiency, and deliver personalized services to users. AI-driven parking systems harness the power of data analytics, machine learning algorithms, and computer vision to optimize various aspects of parking management, including space utilization, pricing strategies, and traffic flow optimization.

One of the key applications of AI in smart car parking stations is predictive analytics, which enables the anticipation of parking demand based on historical data, weather conditions, and events. By forecasting peak demand periods and adjusting parking availability and pricing accordingly, predictive analytics optimize space utilization and minimize congestion, ensuring a seamless parking experience for users.

Moreover, AI-powered computer vision technologies are crucial in automating parking enforcement and monitoring processes. By deploying CCTV (closed-circuit television) cameras and image recognition algorithms, smart parking facilities can detect unauthorized parking, monitor occupancy levels in real-time, and facilitate dynamic pricing schemes based on demand fluctuations.

Machine learning algorithms further enhance the efficiency of smart car parking stations by continuously analyzing data patterns and optimizing parking allocation algorithms. By learning from user behavior, traffic patterns, and environmental factors, these algorithms adapt and improve over time, leading to more accurate predictions and better resource allocation.

The integration of AI in smart car parking stations offers a multitude of benefits, including the following:

- Enhanced user experience: AI-driven parking solutions provide users with real-time information on parking availability, optimal routes, and personalized recommendations, enhancing convenience and reducing search time;
- Improved traffic management: By optimizing parking operations and reducing congestion, AI-driven smart parking systems contribute to smoother traffic flow, thereby reducing carbon emissions and alleviating urban congestion;
- Efficient space utilization: AI algorithms optimize parking space allocation, maximizing utilization rates and minimizing wastage, thereby promoting sustainable urban development and resource efficiency.

However, adopting AI in smart car parking stations also presents several challenges, including concerns regarding data privacy, cybersecurity risks, and the digital divide. Moreover, the upfront costs associated with deploying AI infrastructure and the need for interoperability with existing transportation systems pose implementation challenges for stakeholders.

2 Methods

The first part of this research presents a survey conducted for the Smart Parking application on how successful the application can be in the city where the survey was conducted. The second part describes the parking application, its advantages, and disadvantages.

The opinion of users is important for every new public project. This online survey was conducted among drivers in Tashkent. The survey involved 41 drivers.

The questionnaire consisted of the following ten questions:

- Q1: Do you have to look for many places to park your car, or can you find a place easily the first time?
- Q2: How long will it take you to find a parking place?
- Q3: How often do you use city center parking lots?
- Q4: How often do you have to look for a parking space?
- Q5: Have you changed your plans by abandoning your planned work while searching for a parking place (e.g., entering the market or various entertainment venues)?
- Q6: Have you witnessed an accident in the parking lot?
- Q7: Would you like to use a smartphone application that shows the number of free spaces in parking lots and guides you to a parking spot?
- Q8: Do you want to pre-book a parking space for your car through a smart application?
- Q9: Do you want to download the "Smart Parking" application with the above features to your smartphone?
- Q10: Do you use a GPS navigator? What kind?

The availability of transport telematics in cities provides an opportunity to provide constant information in real time. If the information about the parking lots and the available space is delivered directly to the driver, it is possible to eliminate the losses caused by the search for a place.

Setting up a separate building or off-street parking requires much money. It also affects the design of the city. Parking of cars on the side of the street can cause traffic jams. According to Shoup, a city planner, the free parking on the side of the street leads to more cars on the street and traffic [10].

The lack of Internet access, telematics tools, and intelligent equipment and technologies is considered a drawback in implementing the "Smart Parking" system. However, through the mobile application, we will have the opportunity to receive information about the parking lot in real time, reserve spaces in advance, and pay for the space. The costs of searching for a parking space may be higher than the costs of organizing this system.

3 Results and Discussion

The results also consist of two parts. The first part provides information on the survey results. The second part presents the theoretical scheme of the "Smart parking" application.

First, the authors analyzed the answers of drivers to questions. When the answers were analyzed after the survey, it was found that most drivers (95.1%) have difficulty finding parking for their cars. The driver has to spend the following time to find a parking space (Table 1, part 1).

Table 1 Time spent searching for space and change of plans of drivers while searching for space

No.	Answers	Percentage (%)
Time spent searching for space (*part 1*)		
1	Up to 5 min	14.6
2	6–10 min	51.2
3	11–20 min	26.8
4	More than 20 min	7.3
Change of plans of drivers while searching for space (*part 2*)		
1	Yes, there have been many such cases	24.4
2	Yes, sometimes. In most cases, I find a space, even if it takes time	73.2
3	No, there was no such situation	2.4

Source Created by the authors

Table 1 (*part 1*) shows that, on average, drivers spend 6–10 min searching for a free space. Many of these drivers use the city center car parks all the time or usually very quickly.

Some drivers (24.4%) say they abandon their planned work while searching for a free place. Most of them find a place even if it takes much time (Table 1, part 2).

Simultaneously, traffic jams are observed even more when searching for free space in cities. Various unfortunate incidents can occur because drivers do not pay enough attention to the surroundings while looking for a place. Of all the respondents, 19 said they witnessed accidents in the residential area, and 3 drivers indicated that this incident happened to them (Table 2, part 1).

Table 2 Car accident in the parking and desire to use the mobile application

No.	Answers	Percentage (%)
Car accident in the parking (*part 1*)		
1	Yes, it happened to me	7.3
2	Yes, I witnessed it	46.3
3	No, I have not come across it	46.3
Want to use the mobile application (*part 2*)		
1	Yes	100
2	No	0

Source Created by the authors

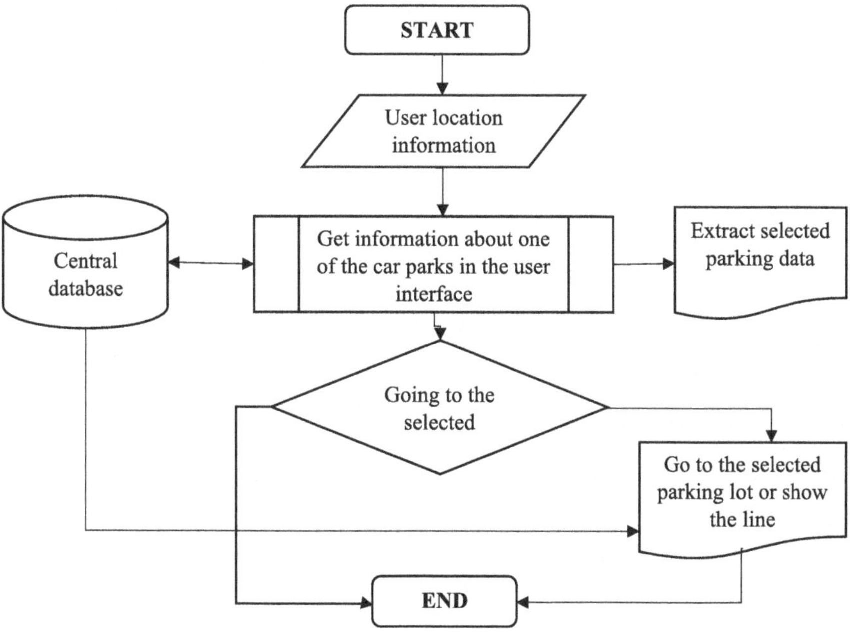

Scheme 1 Parking lot search and directions scheme. *Source* Created by the authors

All drivers (100%) gave positive answers to the questions about downloading and using the "Smart parking" application on a smartphone as a solution to these problems (Table 2, part 2). All drivers use the GPS navigator through a smartphone application.

From the answers to the survey questions, we can say that this mobile application is an effective solution for reducing losses in the search for a parking place and saving time for drivers [11].

The theoretical design of the "Smart parking" application is shown in the following diagrams. Scheme 1 shows the information provided in the driver interface.

Scheme 2 shows the sending of information about the arrival and departure of cars to the central database.

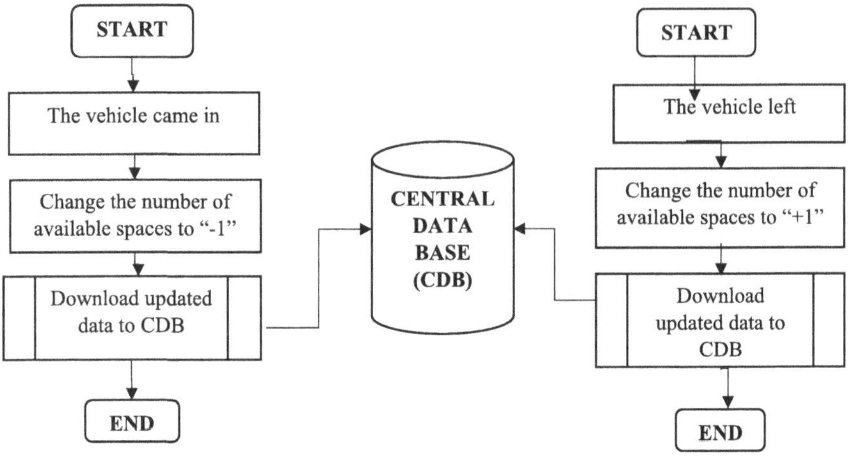

Scheme 2 Scheme of arrival and departure of the car to the parking lot. *Source* Created by the authors

4 Conclusion

The research discusses the pressing issue of parking shortage in urban areas and proposes the integration of AI in smart car parking stations as a solution. It begins by highlighting the challenges of parking space management, congestion, and pollution in cities due to the proliferation of vehicles. Integrating AI technologies (e.g., predictive analytics, machine learning, and computer vision) in smart parking systems is explored, emphasizing their potential to optimize parking operations, enhance user experience, and mitigate urban congestion.

The research presents a comprehensive overview of the benefits and challenges associated with AI-enabled parking solutions, including enhanced user experience, improved traffic management, and efficient space utilization. It discusses the results of a survey conducted among drivers in Tashkent regarding their parking experiences and willingness to adopt a mobile application-based parking system. The survey results indicate a high interest among drivers in using the proposed "Smart Parking" application.

Furthermore, the research provides theoretical designs and schemes for implementing the "Smart Parking" application, including the driver interface and the process of sending information about cars' arrival and departure to the central base. Theoretical analysis and empirical evidence from the survey support the effectiveness of the proposed AI-driven parking solution in reducing parking search time and improving overall efficiency.

Overall, the research presents a compelling case for adopting AI in smart car parking stations to address the challenges of parking space shortage, congestion, and pollution in urban areas. It highlights the transformative impact of AI technologies on parking management and emphasizes the importance of innovative solutions in shaping the cities of tomorrow.

References

1. Kalašová A, Čulík K, Poliak M, Otahálová Z (2021) Smart parking applications and its efficiency. Sustainability 13(11):6031. https://doi.org/10.3390/su13116031
2. Saltzman RM (1994) Three proposals for improving short-term on-street parking. Socioecon Plann Sci 28(2):85–100. https://doi.org/10.1016/0038-0121(94)90008-6
3. Xalilova GX, Samatov RG (2023) Parking search problems and solutions. Dev Innov Sci 4:19–21. https://doi.org/10.5281/zenodo.7839574
4. Ali G, Ali T, Irfan M, Draz U, Sohail M, Glowacz A et al (2020) IoT based smart parking system using deep long short memory network. Electronics 9(10):1696. https://doi.org/10.3390/electronics9101696
5. Shao W, Salim FD, Gu T, Dinh N-T, Chan J (2017) Traveling officer problem: managing car parking violations efficiently using sensor data. IEEE Internet Things J 5(2):802–810. https://doi.org/10.1109/JIOT.2017.2759218
6. Santana Krishnan R, Lakshmi Narayanan K, Bharathi ST, Deepa N, Mathumitha Murali S, Ashok Kumar M et al (2022) Machine learning based efficient and secured car parking system. In: Balas VE, Solanki VK, Kumar R (eds) Recent advances in Internet of things and machine learning. Springer, Cham, Switzerland, pp 129–145. https://doi.org/10.1007/978-3-030-90119-6_11
7. Mangiaracina R, Tumino A, Miragliotta G, Salvadori G, Perego A (2017) Smart parking management in a smart city: costs and benefits. In: Proceedings of SOLI: IEEE international conference on service operations and logistics, and informatics. IEEE, Bari, Italy, pp 27–32. https://doi.org/10.1109/SOLI.2017.8120964
8. Samatov RG, Xalilova GX (2022) Improving efficiency through a program created in python in the organization of smart parking lots. Eurasian J Acad Res 2(13):916–918. https://doi.org/10.5281/zenodo.7472027
9. TJB DD, Subramani A, Solanki VK (2017) Smart city: IOT based prototype for parking monitoring and management system commanded by mobile app. Ann Comput Sci Inf Syst 10:341–343
10. Manville M, Shoup D (2005) Parking, people, and cities. J Urban Plan Dev 131(4):233–245. https://doi.org/10.1061/(ASCE)0733-9488(2005)131:4(233)
11. Xalilova GX (2023) Smart parking mobile application and its effectiveness. Acad Res Educ Sci 4(11):391–398

Creation and Forecasting of Macroeconomic Models Using Information Technologies in the Transport Industry of the Kyrgyz Republic

Elena V. Zenina, Anna I. Kramarenko, Mykash I. Israilov, Maria S. Stoyka, and Oksana A. Sadovskaya

Abstract The creation and forecasting of macroeconomic models are becoming increasingly important in today's world, where the transport industry plays a key role in the country's economic development. Like in many other countries, the transport industry of the Kyrgyz Republic faces certain challenges and problems, including inefficient resource use, insufficient infrastructure, limited budget funds, etc. The use of information technologies to create and forecast macroeconomic models in the transport industry is becoming a necessity for effective resource management and development of the industry. One of the main problems faced in the transport industry of the Kyrgyz Republic is the lack of accurate and reliable macroeconomic models that can consider various factors influencing the industry's development. It is also necessary to forecast the development of transport infrastructure and economic indicators to make informed management decisions. The research aims to develop and apply macroeconomic models using information technologies in the transport industry of the Kyrgyz Republic. The main objectives of this research are (1) to develop new methods and forecasting models considering the transport industry specifics, (2) to use information technologies for collecting, analyzing, and interpreting data, and (3) to analyze the effectiveness of developed models and their applicability for practical use in the transport industry management.

Keywords Model · Information technology · Forecast · Methods · Analysis

E. V. Zenina (✉) · A. I. Kramarenko
International University of Kyrgyzstan, Bishkek, Kyrgyzstan
e-mail: elena.zenina@iuk.kg

M. I. Israilov · M. S. Stoyka
Kyrgyz-Russian Slavic University named after B.N. Yeltsin, Bishkek, Kyrgyzstan
e-mail: ademi_2009_85@inbox.ru

O. A. Sadovskaya
Kyrgyz Aviation Institute named after I. Abdraimov, Bishkek, Kyrgyzstan

S. M. Sultanova et al. (eds.), *Sustainable Development of Transport*, Lecture Notes in Networks and Systems 1366, https://doi.org/10.1007/978-3-031-88846-5_16

JEL codes C1 · C2 · C3 · C4 · D2 · D3 · H1 · H3 · O1 · O5

1 Introduction

The creation and forecasting of macroeconomic models using information technologies largely depends on a high-quality and complete consideration of all statistical and financial results of the transport industry [5] (p. 21).

The creation and forecasting of macroeconomic models using information technologies begins with the study of time series in the EViews statistical package, which makes it possible to solve many problems related to the scientific and financial information analysis, the macroeconomic models and their forecasting creation, etc. [6] (p. 7).

EViews is a statistical package designed primarily for econometric time series data analysis, panel data analysis and modeling, and regression model construction.

Using the econometric package EViews, the authors will analyze statistical series and build models, which will be further used in determining the forecast value for 2024. By comparing the forecast value with the actual level of freight turnover by transport mode, the authors will determine which model most reliably reflects the series' dynamics. This operation will make it possible to predict freight turnover by transport mode in the following periods. It is necessary to keep in mind the "black swans," the influence of which was described by Nassim Taleb [10]. They are bringing extraordinary and difficult-to-predict changes to the current economic and political environment. One of these "black swans" for the Kyrgyz Republic was the COVID-19 pandemic, which is clearly visible in the freight turnover decline by transport mode.

2 Materials and Methods

In this research, the authors used the following statistical methods: testing for stationarity of the time series using the Dickey–Fuller unit root test, visual analysis of autocorrelation using the correlogram, building a regression model using the least squares method to identify dependencies, modeling time series using ARIMA models, and testing the regression model assumptions for autocorrelation of residuals using the Breusch–Godfrey test [2] (p. 71).

The Dickey-Fuller test is a statistical test used to detect the presence of unit roots (a property of non-stationarity) in time series data. It is widely used in fields that analyze data collected over different points in time, such as econometrics, finance, and others. It is one of the most popular tools for investigating the characteristics of time series because the presence or absence of unit roots influences the choice of appropriate methods for modeling and analyzing such data.

A Correlogram, also known as an autocorrelation function (ACF), represents the correlation between the time series values and its lagged ones. This analysis tool is widely used in statistics, econometrics, and time series analysis to identify patterns and structure in data [9] (p. 129).

The Correlogram description includes the following key elements:

- The horizontal X axis displays lag numbers (i.e., the number of periods by which the data lags are behind the current point in time). Typically, lags are indicated by integers starting from zero.
- The vertical Y-axis displays the correlation coefficient values between the time series and their lagged versions. Correlation values can range from -1 to 1, where 1 denotes a perfect inverse correlation, 0 denotes no correlation, and 1 denotes a perfect forward correlation.

Each line or column on a Correlogram represents the correlation between the data and its lagged versions by a specific lag. Typically, several lags are shown on a Correlogram to understand which lags are significant for the analyzed time series.

The lines or columns thickness or length in a Correlogram indicates the correlation strength between the series and their lags. Thicker or longer lines usually indicate a stronger correlation.

Correlograms are often used to identify seasonality, trends, and other time patterns in data. They also help determine the optimal number of lags to use in time series forecasting models.

Building a regression model using the least squares method (LSM) is one of the main tools in data analysis and statistics. This method is used to estimate the parameters of a linear model that describes the relationship between a dependent variable and one or more independent variables [11] (p. 82).

The least squares method is one of the most common and effective methods for estimating the regression model parameters. It is widely used in various fields, including econometrics, sociology, medicine, finance, and others [7] (p. 246).

The development of ARIMA models (Autoregressive Integrated Moving Average) represents a crucial tool for time series analysis and forecasting across diverse domains, including economics, finance, meteorology, and many others [4] (p. 71).

The Breusch-Godfrey test is a statistical test used to control autoregressive conditional heteroscedasticity (ARCH) in regression models. This test is often used in econometrics to assess the presence of autocorrelation in model residuals after regression analysis [3] (p. 129).

The Breusch-Godfrey test is often used in econometrics and financial analytics to control the regression models' assumptions and ensure the parameter estimates adequacy. This test helps researchers and analysts improve the models' quality and obtain more accurate predictions or conclusions [11] (p. 54).

3 Results

As an information base for the macroeconomic model, the authors use the freight turnover indicator in the Kyrgyz Republic in 1994–2023 in million ton kilometers (Fig. 1).

According to Fig. 1, freight turnover in the Kyrgyz Republic, despite fluctuations, has an upward trend. The reccession in 2020 was a consequence of the COVID-19 pandemic, which largely stopped all movement in the country.

Thus, considering that the time series model belongs to the category of statistical models (the dependence of future values on previous ones in the form of an equation), the authors use a linear regression model and an ARIMA autoregression model.

Since the last series value known to us is 2982.9 million ton kilometers in 2023, we will use it as an actual value for comparison with the obtained forecast from the models (Table 1).

The series correlogram showed the linear trend presence. We can notice a gradual decrease in ACF (autocorrelation) and a sharp break in PAC.

Therefore, we assume that the best model is AR(1)—a first-order autoregression model.

Let us construct a trend equation (Y @TREND() c—regression equation) to check the linear trend presence. This operation is performed by using the least squares method, which is the most popular in regression analysis.

A total of 29 observations are presented in the sample. The series value for 2023 is not included, as it will be compared with the model forecasts (Table 2).

$$\text{The resulting equation: GRYZOOBOROT} = 51.395 * t + 1416.84$$

$$R^2 = 87.8\%.$$

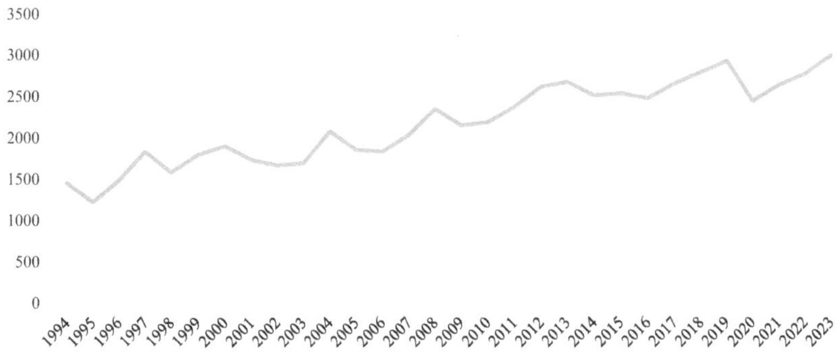

Fig. 1 Freight turnover in the Kyrgyz Republic for 1994–2023, million ton kilometers. *Source* Compiled by the authors based [8]

Table 1 Series correlogram

Sample: 1994–2023						Prob.
Included observations: 29				PAC	Q-Stat	
Autocorrelation	Partial correlation		AC			
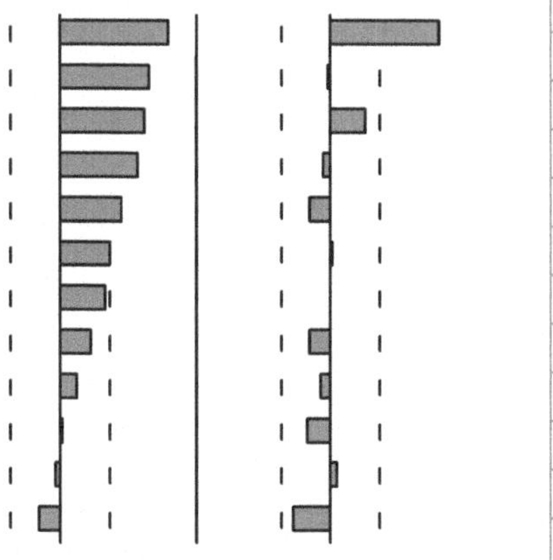		1	0.827	0.827	21.956	0.000
		2	0.679	−0.017	37.286	0.000
		3	0.644	0.276	51.627	0.000
		4	0.589	−0.051	64.109	0.000
		5	0.463	−0.169	72.143	0.000
		6	0.377	0.015	77.703	0.000
		7	0.337	0.004	82.355	0.000
		8	0.240	−0.162	84.821	0.000
		9	0.127	−0.068	85.543	0.000
		10	0.020	−0.187	85.562	0.000
		11	−0.035	0.046	85.624	0.000
		12	−0.158	−0.280	86.951	0.000

Source Compiled by the authors based [1]

Table 2 Equation estimation GRYZOOBOROT @trend C

Dependent variable: GRYZOOBOROT				
Method: least squares				
Sample (adjusted): 1994 2022				
Included observations: 29 after adjustments				
Variable	Coefficient	Std. error	t-Statistic	Prob.
@TREND	51.39517	3.686560	13.94123	0.0000
C	1416.840	60.12598	23.56452	0.0000
R-squared	0.878026	Mean dependent var		2136.372
Adjusted R-squared	0.873508	S.D. dependent var		467.0224
S.E. of regression	166.0999	Akaike info criterion		13.12953
Sum squared resid	744907.7	Schwarz criterion		13.22382
Log likelihood	−188.3782	Hannan–Quinn criter.		13.15906
F-statistic	194.3578	Durbin–Watson stat		1.613234

Source Compiled by the authors based [1]

The adjusted coefficient of determination is 87.3%; the Akaike and Schwartz information criteria are 13.13 and 13.22, respectively.

Upon examining the dynamics of the residuals, the original series for the trend equation, we notice that the residuals aerogenesis, a ward trend, can be observed in the series.

According to the resulting model, the forecast value for 2023 was 2907.3 million ton kilometers.

The second model is the autoregressive integrated moving average model. First, the authors will conduct the Dickey-Fuller test. The null hypothesis of the freight turnover series has a unit root. There are three versions of the test in total, with only a constant and without a constant and trend. Through selection, the test with a constant was chosen. With a lag length of 2, the t-statistic value of augmented test statistic was −5177; critical values were −43743, −0.6032, and −3.381 for the 1, 5, and 10% levels.

The series is stationary as the probability is equal to 0. The Dickey-Fuller statistic is less than the critical values at all significance levels. Considering that the series with a constant and trend is stationary, we use it to determine the structure of the AR model taking into account integration (ARI).

The Breusch-Godfrey LM test is necessary to find the Moving Average (MA) variability in the model. The null hypothesis is considered—there is no autocorrelation in the freight turnover series. The probabilities of f-statistics and chi-square are close to 0. We reject the null hypothesis. MA(1) has 10% probability. All other indicators are insignificant.

$R^2 = 66.1$. Based on the LM test, it was decided to include a first-order moving average model into the model MA(1).

By selecting variabilities using the least squares method on a sample from 1996 to 2022 after 12 iterations, it will be clear that the best model is ARIMA (1;1;1). The analysis was based on 27 observations (after adjustments). All variables (i.e., C, AR(1), and MA(1)) are significant.

$R^2 = 42\%$. The adjusted determination coefficient is 36%. The Akaike and Schwartz information criteria are 13.15 and 13.3, respectively.

The Durbin-Watson statistic is 8.4, indicating no autocorrelation in the model resides. The value above suggests negative dynamics.

The Hannan-Quinn criterion is 1320. This information criterion is used to compare the quality of different models. The lower the criterion, the better the fits.

The mean value of the variable indicates that the valeted indicator is 1.95 over the considered period.

The model includes AR()—the first-order autoregressive component that considers the influence of the value of the time series on the moving average component. It considers the influence of random error in the previous observation on the current value.

The AR(1) provides the autoregressive effect and the first-order moving average effect to model the dynamics of the time series. The statistics indicate a fairly good-quality model.

Based on the obtained model, the authors derive the following equation (Fig. 2):

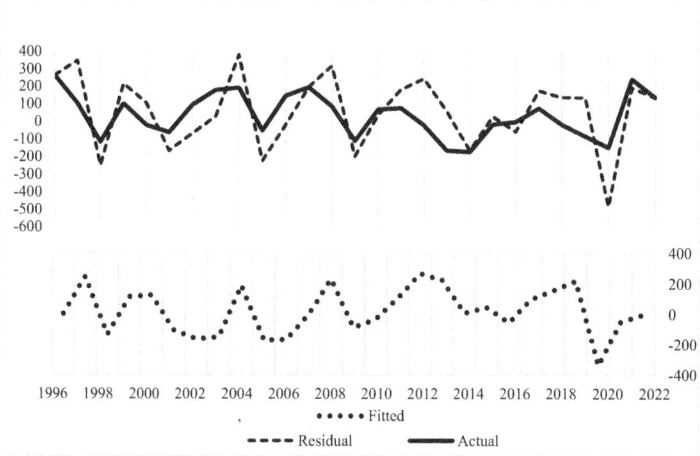

Fig. 2 Dynamics of residuals, original and simulated series for ARIMA (1;1;1). *Source* Compiled by the authors based on [1]

$$D(GRYZOOBOROT) = 47.616 + 0.695 * AR(1) - 0.999 * MA(1).$$

Model residuals are heterogeneous. Let us make a forecast. The obtained forecast for 2023 is 2836.1538 million ton kilometers.

Let us compare the obtained models' actual value. The linear value is closer to the actual value than ARIMA (1;1;1). Therefore, this model is better for Freight Turnover in the Kyrgyz Republic for the period from 1994 to 2022. According to the NSC KR data, the freight in 2023 amounted to 2982 million tons, while the forecasted indicator was 2907.3 million ton kilometers. According to this model, the freight turnover volume will be 2968.8 million ton kilometers in 2024.

4 Conclusion

Based on the methods used, the following conclusions can be drawn.

The Dickey-Fuller test was used to control the time series stationarity. The test results made it possible to determine whether the time series is stationary or requires additional processing to improve forecasts.

Correlogram analysis made it possible to determine the seasonality and autocorrelation presence in the data. This is important for understanding the relationships between time series values and selecting an appropriate forecasting model.

The regression model construction using the least squares method allowed the authors to estimate the regression parameters and obtain an equation that describes the relationship between the dependent and independent variables. Regression analysis can be useful in identifying factors that influence changes in a time series.

ARIMA (Autoregressive Integrated Moving Average) models were used for time series forecasting. They consider autocorrelation and seasonality in the data, making them effective for predicting future values.

The Breusch-Godfrey test was used to control the regression model residuals autocorrelation. It helps assess the model's adequacy and identify the presence of structural changes in the data.

The use of these methods makes it possible to comprehensively analyze time series and construct reliable forecasting models, which is an important tool for making informed management decisions in the transport industry of the Kyrgyz Republic.

The dynamics of cargo turnover in the Kyrgyz Republic are most closely described by the following equation: GRYZOOBOROT = 51.395 * t + 1416.84. The resulting equation relates to statistical models, namely linear regression models. According to the NSC KR, the actual volume of cargo turnover in 2023 amounted to 2982.9 million ton kilometers. The forecast value of the model was 2907.3 million ton kilometers. According to this model, the volume of cargo turnover in 2024 will be 2968.8 million ton kilometers.

References

1. EViews (n.d.) https://www.eviews.com. Accessed 15 Jan 2024
2. Greene WH (2017) Econometric analysis, 8th edn. Prentice Hall, Hoboken, NJ
3. Gujarati DN, Porter DC (2008) Basic econometrics, 5th edn. McGraw-Hill Irwin, New York, NY
4. Hyndman RJ, Athanasopoulos G (2018) Forecasting: principles and practice. OTexts, Melbourne, Australia
5. Khutoretsky AV, Goryushkin AA (2024) Mathematical models and methods: textbook. Lan', St. Petersburg, Russia
6. Kuznetsov IA, Nagaeva IA (2024) Fundamentals of mathematical modeling and numerical methods: textbook, 2nd edn. Lan', St. Petersburg, Russia
7. Montgomery DC, Peck EA, Vining GG (2015) Introduction to linear regression analysis. John Wiley & Sons, Hoboken, NJ
8. National Statistical Committee of the Kyrgyz Republic (n.d.) Freight turnover in the Kyrgyz Republic for the period 1994–2023. https://stat.kg/ru/opendata/. Accessed 15 Jan 2024
9. Shumway RH, Stoffer DS (2017) ARIMA models. In: Time series analysis and its applications: with R examples. Springer, Cham, Switzerland, pp 75–163. https://doi.org/10.1007/978-3-319-52452-8_3
10. Taleb NN (2015) The black swan: the impact of the highly improbable, 2nd edn (Transl. from English). CoLibri, Moscow, Russia (Original work published 2008)
11. Wooldridge JM (2015) Introductory econometrics: a modern approach, 6th edn. Nelson Education, Toronto, Canada

Logistics in the Digital World: Transition to Logistics 4.0

Irina F. Zhuckovskaya, Olga B. Yares, Marina V. Krasnova, and Maria M. Markhaichuk

Abstract The Fourth Industrial Revolution has resulted in new technologies impacting the development of the real economy and virtually every industry and sector. It also transformed business models into digital business models. Logistics did not escape this either. This research aims to analyze the evolution of logistics under the influence of industrial revolutions, current trends in digitalization, and its transition to Logistics 4.0. The authors applied analysis, synthesis, induction, deduction, cause-and-effect relationships, comparison, and description during the research. The authors identified the stages of logistics development—from Logistics 1.0 to Logistics 4.0; the characteristics of these stages are provided. The authors also established that the logistics sector has developed over the past decades due to multidirectional factors. During the COVID-19 pandemic, the introduction of digital technologies and robotization of logistics has accelerated for several years ahead, resulting in the development of Logistics 4.0. The authors defined Logistics 4.0, which introduces novel elements. It is described as a customer-oriented business model within value chains, utilizing an integrated approach to organizing and managing material and information flows. This model incorporates digital technologies that align with the scientific and technological demands of Industry 4.0. The results obtained are of interest for further research into Logistics 4.0 from the point of view of its conceptualization and assessment of its effectiveness in practice.

Keywords Logistics · Logistics 4.0 · Digital technologies and innovations · Value chain · Industry 4.0

I. F. Zhuckovskaya (✉) · O. B. Yares · M. V. Krasnova · M. M. Markhaichuk
Vladimir State University, Vladimir, Russia
e-mail: ateya33@mail.ru

O. B. Yares
e-mail: yaresolga@inbox.ru

M. V. Krasnova
e-mail: krasnovam772@mail.ru

M. M. Markhaichuk
e-mail: m.marhaychuk@gmail.com

S. M. Sultanova et al. (eds.), *Sustainable Development of Transport*, Lecture Notes in Networks and Systems 1366, https://doi.org/10.1007/978-3-031-88846-5_17

JEL code R4

1 Introduction

The Fourth Industrial Revolution, also known as Industry 4.0, involves the extensive application of solutions designed to enhance technical and technological infrastructure, drive innovation, and integrate new generations of digital technologies. These advancements aim to improve economic performance and boost business competitiveness.

Digital transformation of production processes and intensive technological changes led to a digital revolution in cargo handling and management practices supply chains.

The digital transformation accelerated during the COVID-19 pandemic when the governments of most countries worldwide imposed strict restrictions on movement. During this period, logistics companies realized that traditionally used methods and tools could not provide sufficient flexibility in supply chains according to the needs and requirements of the market. To improve the performance of supply chains and increase customer satisfaction in timeliness, quality, and delivery locations, the logistics sector has become even more active in integrating technologies and principles of Industry 4.0 into business processes.

This research aims to analyze the evolution of logistics under the influence of industrial revolutions, the trends in its digitalization at present, and its transition to Logistics 4.0. In the research, the authors tested the hypothesis that changes in the technological landscape and adopting Industry 4.0 technologies have transformed trade conditions. Moreover, during and after the COVID-19 pandemic, the digitalization of the logistics sector has accelerated, marking the emergence of a new era—Logistics 4.0.

2 Materials and Method

The development of Industry 4.0 and digital transformation has provided new opportunities for various sectors of the economy at various stages of the value chain. Many researchers considered the impact of advanced technologies on the economy, including logistics. Particularly, some works published in the pre-pandemic period were devoted to the analysis of smart products [10] or smart manufacturing [21], global supply chains in Industry 4.0 [9, 18], and organizational factors contributing to the use of intelligent supply chains [3, 8, 26].

During this period, separate works reflecting the formation of separate smart subsystems in logistics that meet the requirements of Industry 4.0 also appeared. Particularly, G. F. Frederico, J. A. Garza-Reyes, A. Anosike, and V. Kumar introduced the term "Supply Chain 4.0" [7] and identified the following technologies as the technological lever of Industry 4.0: "Internet of Things (IoT), smart products (SP), smart machines (SM), cybersecurity and blockchain (CSB), artificial intelligence (AI), big data analysis (BDA), cloud technologies (CT), machine-to-machine communication (M2M), etc." [16].

Some researchers point to the great potential of particular digital technologies and their positive impact on improving the efficiency of decision-making [4] and the sustainability of logistics business [1], supply chain management, transparency [26], and the need to form a unified digital ecosystem.

The acceleration of digitalization during the COVID-19 pandemic led to revolutionary transformations in the field of logistics and the emergence of the terms Logistics 4.0 [2, 5, 6, 12] and intralogistics [22, 25]. However, much of the existing work primarily consists of reviews of research and publications that discuss the relationship between logistics and Industry 4.0 technologies, as well as occasional references to the term Logistics 4.0. Despite this, a clear and precise definition of Logistics 4.0 has yet to be established.

3 Results

To understand revolutionary transformations in logistics, it is necessary to consider the previous stages of its development.

The history of logistics has its roots in the distant past when it was developed as a military discipline for organizing the work of rear units to provide troops with everything they needed.

The first economic breakthrough is associated with the transition from manual production to factory production.

Mass production of inexpensive products began to develop during the Second Industrial Revolution [24]. The introduction of electricity as a source of energy, the invention of the telegraph, and the expansion of railroads helped reduce the time required to travel, transport goods, and transmit information.

However, until the mid-twentieth century, logistics was widely used only in the military sector and shifted towards business activities only in the 1960s.

In 1964, logistics was first presented as a business process and was called business logistics, which considered the issues of optimizing the transportation and movement of goods. Thus, Logistics 1.0 was engaged in optimizing three "Ps" [19]: place, period and pace, and pattern.

By the end of the 1960s, Industry 2.0 realized the importance of mass production, which required the introduction of automation of cargo handling and loading and unloading processes and marked the transition to Logistics 2.0.

In the 1980s, companies began to closely cooperate in the field of managing and coordinating material flows within their organizations and outside them; the term "supply chain management" appeared.

Sheffi and Klaus defined supply chain management as the logistics of the four "Ps," adding "process coordination or partnerships management" to the previous three [19].

The third industrial revolution marked the transition to Industry 3.0—from mechanical and analog technologies to digital electronics (widespread computing, penetration of the Internet into all areas of life, mass use of personal portable communication devices, etc.).

Since the 1990s, logistics has penetrated all sectors of the economy. This stage is characterized by the introduction of automated warehouse complexes, the intensification of container transportation of goods, and the use of uniform container and packaging standards in logistics systems. The functions of manufacturing enterprises and their partners are being combined into a logistics chain: procurement—production—sales. The emergence of software in the form of various logistics management systems in the twenty-first century marked the transition to Logistics 3.0 and the introduction of the fifth "P"—"pliancy." Additionally, the volume of logistics services outsourced to third-party logistics providers has been steadily increasing worldwide.

Thus, by comparing the transformation of the economy and logistics, we can conclude that the evolution of logistics did not occur simultaneously with the industrial revolutions but followed them (Table 1).

Klaus Schwab outlined the concept of the Fourth Industrial Revolution in his 2016 book of the same name [17]. Industry 4.0 involves the massive introduction of cyber-physical systems (CPS), existing in physical and cyberspace, into the production and servicing of human needs.

According to experts, the volume of the global logistics market in 2023 amounted to $9.66 trillion. By 2028, it is projected to reach $13.77 trillion, increasing by an average of 6.23% [16].

By its structure, the logistics sector includes four main segments: transport (land, water, and air transport) and forwarding, warehousing services, inventory management and administration, and supplies. The largest market share is traditionally occupied by transport and forwarding services (55%) and warehousing services (25.6%). The third-party logistics sector (i.e., logistics outsourcing) is also growing.

Over the past 15 years, international logistics has evolved due to changing technological frameworks and multidirectional trends, an increased frequency of catastrophes, and various crises. These factors have contributed to greater instability and

Table 1 Timelines of industrial revolutions and logistics

Industry 1.0	Industry 2.0	Industry 3.0	Industry 4.0
the eighteenth century—1st half of the nineteenth century	2nd half of the nineteenth century—mid-twentieth century	since 1960s—early 2000s	since 2010s (2016)
Logistics 1.0	Logistics 2.0	Logistics 3.0	Logistics 4.0
1960–1970s (1964)	1980s	since 1993–1997	since 2020s

Source Compiled by the authors

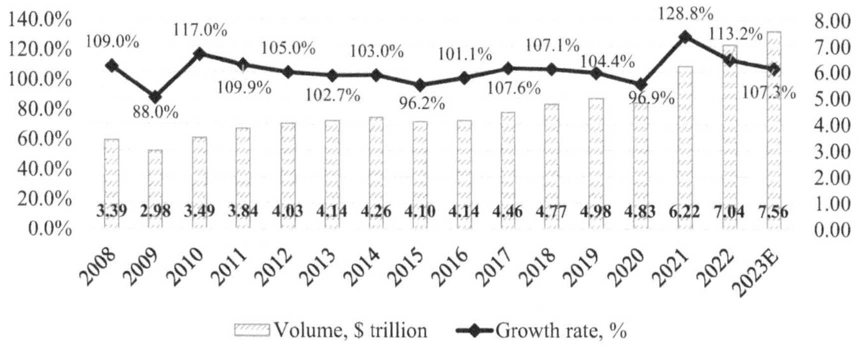

Fig. 1 Dynamics of the volume of the global TLS market ($ trillion) and its growth rate (% compared to the previous year). *Source* Compiled by the authors based on [11]

uncertainty, resulting in fluctuating periods of growth and decline. This is particularly evident in the volatility within the transport and logistics services (TLS) market (Fig. 1).

At the beginning of the twenty-first century, a significant increase in trade turnover contributed to developing the transport and logistics services market. Against the backdrop of the global economic crisis (2009), the drop in production volumes, and the reduction in global exports and imports, there was a decrease in demand for logistics services. The recovery period that began at the end of 2009 led to increased supplies and a market revival. However, the large-scale sanctions introduced in 2014 and a significant drop in global fuel and raw material prices triggered a crisis in the global economy, including the logistics sector. The subsequent recovery that started in 2016 was abruptly halted by the crisis caused by the COVID-19 pandemic.

In the early months of the COVID-19 pandemic, there were isolated supply chain disruptions (especially in the transportation of medical supplies and personal protective equipment). However, large supply chains proved to be more resilient in the crisis, allowing them to play a significant role in economic recovery. This applies, for example, to the trade in parts for the production of passenger cars, where the reduction in volumes was significantly smaller compared to the trade in passenger cars themselves.

Before the COVID-19 pandemic, logistics implemented digital innovation much slower than other industries, as noted at the 2016 World Economic Forum. However, the pandemic has accelerated these processes. For example, the results of a survey of companies conducted by McKinsey in 2020 showed that the implementation of innovation and digital technologies during the pandemic in terms of its indicators is "years ahead of the indicators that were at the time of previous surveys, and even more so in developed Asia than in other regions" [13].

The logistics industry began to recover from the extreme impacts of the COVID-19 pandemic in 2021. However, in 2022, it faced new challenges: sharp increases in commodity and other energy prices, record-high inflation (for many countries), and fluctuations in demand (partly caused by low consumer spending). Difficulties have also arisen when transporting goods by almost all transport types, including lengthening routes, rising tariffs, and insurance premiums. According to The Logistics World [23], Since the second half of 2022, four major problems have emerged in the global economy, including logistics: rising costs, supply chain disruptions, labor shortages, and heightened competition.

According to a survey conducted by Oxford Business among executives from 127 enterprises, geopolitical risks—stemming from military conflicts in Ukraine and the Middle East and the potential for conflict involving South Korea, Russia, China, and Taiwan—are expected to negatively impact business operations and supply chains. Thirty-six percent of respondents share this concern [15].

Rising costs and risks have led the logistics sector to restructure itself and implement digital technologies even more actively. According to the McKinsey survey, 90% of supply chain executives surveyed have invested in digital supply chain management technologies since 2021, and 80% intend to continue investing in the future to support efficiency [14]. Gartner also found that 83% of companies prioritize customer satisfaction by improving the customer experience as part of their digital supply chain management tactics. Robotization of warehouses is also growing. According to Gitnux's findings, the global warehouse automation market will approach $30 billion due to optimal support for rapid manufacturing [20].

Thus, due to the intensive use of digital technologies and Industry 4.0 innovations, the transition from Logistics 3.0 to Logistics 4.0 is taking place much faster.

Most Industry 4.0 technologies are used in the logistics sector, the key ones being the IoT, AI, big data, and others (Fig. 2).

The introduction of digital technologies increases the efficiency and elasticity of logistics. Elastic logistics allows businesses to quickly respond to changes in demand. During peak seasons, companies can increase their production, transportation, and warehouse capacities to meet growing customer needs and orders. When

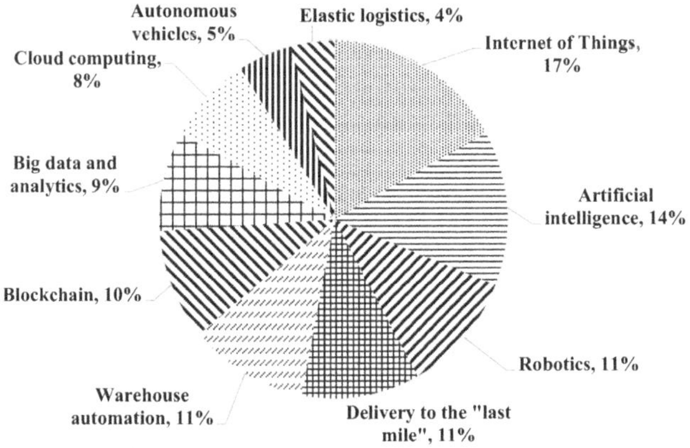

Fig. 2 The best innovations and trends in the logistics industry (%). *Source* Compiled by the authors based on StartUs [20]

demand decreases, they can reduce their scale to avoid unnecessary costs. Elastic logistics makes it possible to solve problems of overstocking in warehouses and reduce transportation costs and the risk of obsolescence of assets.

However, despite the accelerated adoption of digital technologies, businesses face certain challenges because they cannot fully provide connected interactions with users. The reasons are the existence of many outdated codes and systems, disparate data, and a lack of qualified personnel. Additionally, more than half of companies lack risk management associated with the implementation of generative AI [14]. There are also problems of inequality in the implementation of advanced Logistics 4.0 technologies among different countries.

However, unlike the previous three stages, where logistics evolved after the industrial revolutions, the fourth stage is developing almost simultaneously with the revolution.

Having analyzed the definitions of logistics given by various researchers [2, 4, 6, 12], the authors found out that they do not consider the features emerging in the global market of logistics services. The changing geopolitical landscape, the emergence of a multipolar world, shifts in logistics routes, the expansion of national fleets for container and tanker transport, and the growing demand for logistics services (with the share of 3PL services in the global TLS market rising to 20% over the past five years) have all contributed to heightened competition in the logistics services market. This has created a pressing need for greater customer focus in logistics and further integration of Industry 4.0 technologies.

Based on the above, the authors defined Logistics 4.0 as a customer-oriented business within value chains, utilizing an integrated approach to organizing and managing material and information flows, and incorporating digital models that align with the scientific and technological requirements of Industry 4.0.

The key features of Logistics 4.0 include the following:

- The digitization of clients, suppliers, other partners, business processes, products, and services;
- The formation of new business models focused on the entire value chain;
- Building the digital ecosystem around the company;
- Transforming supply chains into value chains;
- Omnichannel interaction with clients and partners;
- Customer focus;
- Flexibility and elasticity, etc.

The formation of new business models in the context of the ongoing digital transformation obliges 3PL and 4PL providers to shift their focus from traditional management business models to business models focused on the entire value chain. Moreover, it allows for the transition to the 5PL level.

4 Conclusion

Thus, the working hypothesis is proven. The analysis carried out during the research showed that the change in technological structures contributes to the evolution of logistics, and the introduction of innovations and technologies of Industry 4.0 leads to the emergence of a new era in logistics—Logistics 4.0.

The theoretical research allowed the authors to formulate the definition of Logistics 4.0, which focuses on the transition from optimizing supply chains and maintaining flow flexibility to value chains and customer focus. According to the authors, such a transition will allow companies to effectively position themselves within value chains by analyzing the competencies of all organizations included in this chain and customer needs.

References

1. Abdul Rahman N, Lirn T, Hamid A, AlKalbani K (2023) Logistics business sustainability incorporating national logistics strategy and industry revolution 4.0. Oper Supply Chain Manag, 16(3), 341–351. https://doi.org/10.31387/oscm0540394
2. Albrecht T, Baier M-S, Gimpel H, Meierhöfer S, Röglinger M, Schlüchtermann J, Will L (2023) Leveraging digital technologies in logistics 4.0: Insights on affordances from intralogistics processes. Inf Syst Front, 26, 7550774. https://doi.org/10.1007/s10796-023-10394-6
3. Ali SB (2022) Industrial revolution 4.0 and supply chain digitization. South Asian J Soc Rev, 1(1), 21–41. https://doi.org/10.57044/SAJSR.2022.1.1.2205
4. Barreto L, Amaral A, Pereira T (2017) Industry 4.0 implications in logistics: An overview. Procedia Manufacturing 13:1245–1252. https://doi.org/10.1016/j.promfg.2017.09.045
5. Benayoune A, Hamid AA, Rahman NSFA, Kalbani KA, Slimi Z (2021) Logistics 4.0 skills requirements: Evidence from a developing country. Can J Bus Inf Stud, 4(2), 24–36. Retrieved from https://tinyurl.com/mr4cuz2d (Accessed 20 February 2024)
6. Bugarčić FŽ, Mijušković VM, Aćimović S (2024) Innovation and new technologies as determinants of logistics 4.0. Politická ekonomie [Political Economy], 72(1), 102–121. https://doi.org/10.18267/j.polek.1422
7. Frederico GF, Garza-Reyes JA, Anosike A, Kumar V (2020) Supply Chain 4.0: Concepts, maturity and research agenda. Supply Chain Manag, 25(2), 262–282. https://doi.org/10.1108/SCM-09-2018-0339
8. Gupta S, Drave VA, Bag S, Luo Z (2019) Leveraging smart supply chain and information system agility for supply chain flexibility. Inf Syst Front 21(3):547–564. https://doi.org/10.1007/s10796-019-09901-5
9. Han L, Hou H, Bi ZM, Yang J, Zheng X (2021) Functional requirements and supply chain digitalization in industry 4.0. Inf Syst Front. https://doi.org/10.1007/s10796-021-10173-1
10. Kahle JH, Marcon É, Ghezzi A, Frank AG (2020) Smart Products value creation in SMEs innovation ecosystems. Technol Forecast Soc Chang 156:120024. https://doi.org/10.1016/j.techfore.2020.120024
11. Research MA (2023). Logistics outsourcing market: Results for 2022, assessment for 2023 and forecast until 2026. Retrieved from https://ma-research.ru/research/item/395-rynok-logisticheskogo-autsorsinga-itogi-2022-g-otsenka-2023-g-i-prognoz-do-2026-g.html (Accessed 10 March 2024)
12. Malagon-Suarez CP, Orjuela-Castro JA (2023) Challenges and trends of the logistics 4.0. Ingeniería, 28(SI), e18492. https://doi.org/10.14483/23448393.18492

13. McKinsey, Company. (2020). How COVID-19 has pushed companies over the technology tipping point—and transformed business forever. Retrieved from https://www.mckinsey.com/capabilities/strategy-and-corporate-finance/our-insights/how-covid-19-has-pushed-companies-over-the-technology-tipping-point-and-transformed-business-forever (Accessed 20 February 2024)
14. McKinsey, Company. The state of AI in 2023: Generative AI's breakout year. Retrieved from https://www.mckinsey.com/capabilities/quantumblack/our-insights/the-state-of-ai-in-2023-generative-ais-breakout-year (Accessed 20 February 2024)
15. Overvest M (n.d.). Logistics statistics 2024—21 key figures. *Procurement Tactics*. Retrieved from https://procurementtactics.com/logistics-statistics/ (Accessed 20 February 2024)
16. Research and Markets (2023) Global Logistics Market Report and Forecast 2024–2032. Retrieved from https://www.researchandmarkets.com/reports/5775157/global-logistics-market-report-forecast (Accessed 10 March 2024)
17. Schwab K (2016) The fourth industrial revolution. Geneva, Switzerland: World Economic Forum
18. Shao X-F, Liu W, Li Y, Chaudhry HR, Yue X-G (2021) Multistage implementation framework for smart supply chain management under industry 4.0. Technol Forecast Soc Chang, 162, 120354. https://doi.org/10.1016/j.techfore.2020.120354
19. Sheffi Y, Klaus P (1997) Logistics at large: jumping the barriers of the logistics function. In proceedings of the twenty-sixth annual transportation and logistics educators conference (pp 1–26), Chicago, IL: Ohio State University. Retrieved from https://www.researchgate.net/publication/266603079_Logistics_at_Large_Jumping_the_Barriers_of_the_Logistics_Function (Accessed 20 February 2024)
20. StartUs Insight (2024) Explore the Top 10 Logistics Trends in 2024. Retrieved from https://www.startus-insights.com/innovators-guide/reverse-logistics-trends/ (Accessed 20 February 2024)
21. Tabim VM, Ayala NF, Frank AG (2021) Implementing vertical integration in the industry 4.0 journey: Which factors influence the process of information systems adoption? Inf Syst Front. https://doi.org/10.1007/s10796-021-10220-x
22. Taş U (2023) Case study of intralogistics in the framework of logistics 4.0. Int J Automot Sci Technol, 7(1), 18–24. https://doi.org/10.30939/ijastech..1215381
23. The Logistics World International Ltd. (n.d.). Worldwide Courier, cargo and logistics service. Retrieved from https://thelwi.com/ (Accessed 20 February 2024)
24. Wang Y, Ma H-S, Yang J-H, Wang K-S (2017). Industry 4.0: A way from mass customization to mass personalization production. Advances in Manufacturing, 5(4), 311–320. https://doi.org/10.1007/s40436-017-0204-7
25. Winkelhaus S, Grosse EH, Glock CH (2022) Job satisfaction: An explorative study on work characteristics changes of employees in Intralogistics 4.0. J Bus Logist, 43(3), 343–367. https://doi.org/10.1111/jbl.12296
26. Zhu XN, Peko G, Sundaram D, Piramuthu S (2021) Blockchain-based agile supply chain framework with IoT. Inf Syst Front 24:563–578. https://doi.org/10.1007/s10796-021-10114-y

Infrastructure Approach to Ensuring Sustainable Development of Transport and Logistics Systems Through Digital Technologies

Elena G. Popkova

Abstract The research explores the significance of infrastructure, including digital technologies, for the sustainable development of transport and logistics systems. Based on the experience of the top 20 countries with the most sustainable transport and logistics systems in 2023, the author developed a model to analyze the impact of structural elements of transport infrastructure on the resilience of logistics systems. Using this model, the author concludes that infrastructure, particularly digital technologies, is critically important for the sustainable development of transport and logistics systems amid the Fourth Industrial Revolution. In light of this conclusion, the author proposed a new "Infrastructure Approach" to ensure the sustainable development of transport and logistics systems based on digital technologies. This approach offers new tools for managing transport logistics, especially suited to the conditions of the Fourth Industrial Revolution: fostering economic openness for globalization, strengthening the marketing power of national brands, and increasing the prevalence and engagement with digital technologies in companies. The infrastructure approach will enable countries with less favorable geographic locations to integrate into global trade routes. It will also allow countries with intergovernmental trade agreements to overcome the moral and physical obsolescence of their transport and logistics systems, thereby stimulating the flow of goods and production factors across their territories. Countries adopting the infrastructure approach will be positioned to achieve sustainable development of their transport and logistics systems in the context of the Fourth Industrial Revolution.

Keywords Infrastructure · Sustainable development · Transport resilience · Transport logistics · International trade · Digital technologies

JEL codes N70 · O32 · R41

E. G. Popkova (✉)
Armenian State University of Economics, Yerevan, Armenia
e-mail: elena.popkova@asue.am

S. M. Sultanova et al. (eds.), *Sustainable Development of Transport*, Lecture Notes in Networks and Systems 1366, https://doi.org/10.1007/978-3-031-88846-5_18

1 Introduction

In today's environment of emerging global trade routes, transport and logistics systems play a crucial role in determining the feasibility, attractiveness, and efficiency of trade routes across each country's territory. The current phase of globalization frames the resilience of transport and logistics systems through their contribution to the growth of international trade. This interpretation is shaped by several factors.

First, access to advanced transport logistics is essential to maintain a favorable trade balance, realize the export potential of national industries, mitigate the risk of production crises through market diversification, and accelerate economic growth via increased business activity. A country's transport and logistics system can be considered resilient if it meets leading global standards and ensures high-speed delivery of goods across a broad range of destinations with minimal transport risks (e.g., risk of damage, loss of goods, or delivery delays) and costs.

Second, import-friendly transport logistics are required for a nation to achieve efficient resource use in its economy and solid global competitiveness for its products by engaging in the international division of labor. In this context, a country's transport and logistics system should be deemed resilient if it enables the import of finished goods, resources, materials, and components from various, even distant, regions. The transportation conditions must be transparent and predictable; the logistics routes should be reliable and stable, supported by all countries through which they pass.

Third, to leverage its geographic position and integration into global trade routes, avoid trade isolation, facilitate beneficial flows of educational and skilled labor migration, and attract the desired number of foreign tourists, a country needs transport logistics that are consumer-friendly, comfortable, and safe for goods and passengers. From this perspective, a country's transport and logistics system can be considered resilient if it promotes the country's central role in the global movement of goods and production factors and allows it to benefit from its transit through its territory.

However, the problem is that the current approach to state regulation of transport and logistics system development provides insufficient support for their resilience, representing a serious shortcoming. The approach adopted by most countries assumes that a nation's attractiveness for routing global trade paths across its territory is predetermined by its geographic location. As a result, countries with less favorable geographic conditions (e.g., remote locations, limited engagement in globalization by neighboring countries, or harsh climates) struggle to achieve sustainable development in their transport and logistics systems.

Moreover, this approach relies heavily on intergovernmental trade agreements, where national regulators often fail to take proactive measures to modernize their transport and logistics systems, expecting global trade routes to pass through regardless of these systems' conditions. This approach instead leads to moral and physical degradation of transport and logistics infrastructure, limiting the flow of goods and production factors through these systems and ultimately reducing their resilience.

This research seeks to address the stated problem by developing an alternative approach to state regulation in developing transport and logistics systems. The foundation of this new approach involves strengthening infrastructure and modernizing transport and logistics systems through integrating digital technologies. The research aims to determine the significance of infrastructure, including digital technologies, for the sustainable development of transport and logistics systems.

2 Literature Review

This research is built on the fundamental framework of the sustainable development concept in transport and logistics systems. According to this concept, recognized scientific criteria for the resilience of these systems include a country's integration into global trade routes to support exports and promote imports [8]. as well as enabling the inflow and transit of international flows of finished goods and production factors, including educational [5] and labor resources [2–4, 10].

A comprehensive literature review culminated in a highly detailed infrastructure support system for transport logistics, which, in the current context of the Fourth Industrial Revolution, includes the following structural elements:

- Availability of digital professionals, forming the workforce base for the modernization of transport and logistics systems [12];
- Financial support for innovations, particularly digital ones, defining the investment needed for modernizing transport and logistics systems [13, 16];
- Economic openness to globalization and exports, setting the intensity of foreign trade operations [1, 7];
- Marketing power of national brands, determining their export potential [11];
- Prevalence and integration of digital technologies in companies, enabling the automation of production and distribution processes related to transport and logistics services [15];
- Diffusion of advanced telecommunications, allowing for the online provision of transport and logistics services [6];
- Social engagement in the Fourth Industrial Revolution, reflecting the level of development of the information society, which influences the appeal of online transport and logistics services [14].

Thus, existing sources and published research thoroughly explain the composition of infrastructure support for transport logistics, particularly emphasizing digital technologies in the Fourth Industrial Revolution. However, the impact of these structural elements, including digital technologies, on the resilience of transport and logistics systems remains underexplored and unclear. This research is conducted to fill this gap in the literature.

3 Materials and Methods

Regression analysis is employed to determine the contribution of each identified structural element of infrastructure, including digital technologies, to the resilience of transport and logistics systems. Using this method, the author studied the relationship between the resilience of transport and logistics systems (represented by "Transport services" (Strlg) as assessed by WITS [17]) and the structural elements of the transport logistics infrastructure (as assessed by IMD [9]):

- "Digital/Technological skills" (i_1) as an indicator of digital workforce availability;
- "Total expenditure on R&D" (i_2), percentage of GDP, representing financial support for innovation;
- "Exports of goods – growth" (i_3), percentage change, based on $ values, reflecting economic openness to globalization from the perspective of export growth rate;
- "Image abroad or branding" (i_4), indicating the marketing strength of national brands;
- "Digital transformation in companies" (i_5), showing the prevalence and integration of digital technologies within companies;
- "Mobile broadband subscribers, 5G market" (i_6), % of the mobile market, reflecting the diffusion of 5G telecommunications;
- "E-Participation" (i_7), representing social engagement in the Fourth Industrial Revolution.

To illustrate leading international practices, the research sample includes the top 20 countries with the most resilient transport and logistics systems. This sample was formed based on the highest Strlg values in 2022 (the latest data available in 2023) (Table 1).

In this research, the author identified structural elements of the transport logistics infrastructure system that are positively associated with its resilience. To establish the prospects for the sustainable development of transport and logistics systems amid the Fourth Industrial Revolution, the author projected the potential impacts of achieving the maximum values observed in the sample for the top 20 countries with the most resilient transport and logistics systems.

4 Results

As a result of regression analysis on the data in Table 1, the authors developed a model to illustrate the impact of structural elements of transport logistics infrastructure on the resilience of transport logistics systems:

$Strlg = 13.52–3.84i_1\text{-}3.76i_2 + 0.11i_3 + 3.86i_4 + 10.65i_5\text{-}0.20i_6\text{-}37.19i_7$, $R^2 = 0.6964$ (1).

Model (1) shows that infrastructure determines 69.64% of the resilience of transport logistics under the conditions of the Fourth Industrial Revolution. The structural

Table 1 Transport logistics infrastructure and its sustainability in 2023

Country	Strlg, % of commercial services exports	i_1, scores 1–10	i_2, % of GDP	i_3, %	i_4, scores 1–10	i_5, scores 1–10	i_6, % of the mobile market	i_7, shares of 1
Chile	49.28	7.00	0.00	-3.67	6.83	5.33	18.33	0.69
China	39.90	7.74	0.00	-4.64	6.81	6.87	43.60	0.86
Denmark	68.31	8.41	2.81	3.67	8.36	7.73	11.94	0.89
Greece	49.81	6.28	1.45	-4.31	6.25	5.72	22.61	0.61
Hungary	28.54	5.46	1.64	6.12	3.76	5.22	12.77	0.51
Iceland	29.58	8.12	2.81	-10.56	8.12	7.58	11.84	0.80
Kazakhstan	60.75	5.78	0.13	-7.16	6.80	6.40	2.44	0.81
Korea Rep	37.33	6.18	4.93	-7.51	8.13	6.90	39.64	0.94
Lithuania	54.90	8.69	1.11	-8.27	6.44	7.03	5.69	0.55
Mongolia	52.61	6.00	0.00	21.10	3.96	5.79	0.00	0.60
Norway	45.29	7.33	1.94	-30.38	6.84	6.71	36.57	0.69
Poland	30.10	6.35	1.44	5.82	5.06	5.95	23.70	0.65
Qatar	64.66	8.14	0.68	-25.66	8.29	7.68	66.11	0.38
Romania	27.41	6.74	0.47	4.05	5.36	5.79	8.19	0.63
Singapore	34.91	7.76	0.00	-7.67	7.86	6.88	19.39	0.98
Slovak Republic	40.53	7.18	0.93	8.24	4.77	5.03	10.59	0.47
Slovenia	30.95	7.00	2.14	4.76	6.20	5.47	23.10	0.75
Turkey	40.82	6.20	1.13	0.63	4.47	5.63	0.00	0.78
Austria	26.96	6.00	3.19	5.48	7.47	5.48	31.03	0.77
Germany	26.45	5.54	3.13	0.75	7.86	4.70	13.50	0.73

Source Developed by the author based on IMD [9] and WITS [17]

elements positively associated with logistics resilience were identified as i_3, i_4, and i_5. The patterns for strengthening resilience in transport logistics systems as these key infrastructure elements improve are as follows:

- Increasing economic openness to globalization by accelerating export growth by 1% raises the share of transport and logistics services in the structure of commercial service exports by 0.11%;
- Enhancing the marketing power of national brands by one point leads to an increase in the share of transport and logistics services in commercial service exports by 3.86%;
- Increasing the prevalence and integration of digital technologies within companies by 1 point boosts the share of transport and logistics services in commercial service exports by 10.65%.

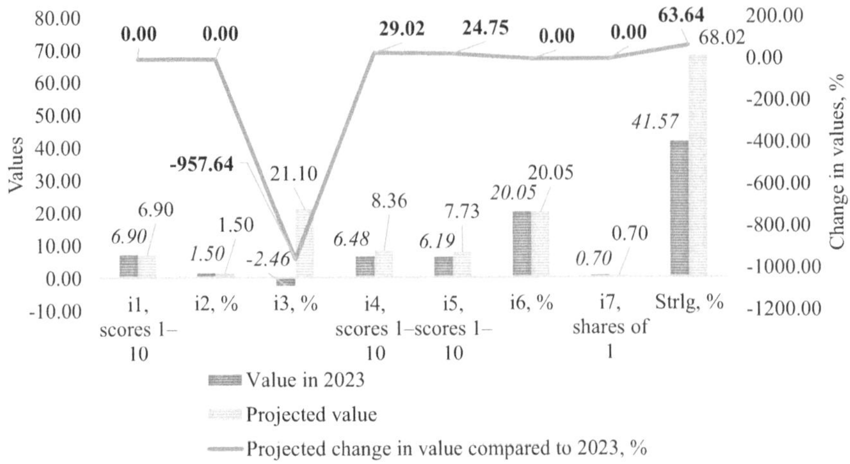

Fig. 1 Perspective of sustainable development of transport and logistics systems in the conditions of the Fourth Industrial Revolution. *Source* Developed by the author

Based on model (1), the author developed an infrastructure approach to ensure the sustainable development of transport and logistics systems through digital technologies, characterized by managing system resilience through targeted improvements in the identified structural elements of transport logistics infrastructure.

Transitioning to this new infrastructure approach offers a promising outlook for the sustainable development of transport and logistics systems in the Fourth Industrial Revolution. A forecast of the potential impact of achieving the maximum sample values for i_3, i_4, and i_5, on average, for the top 20 countries with the most resilient transport and logistics systems is shown in Fig. 1.

The forecast in Fig. 1 shows that, in the context of the Fourth Industrial Revolution, the following results can be expected among the top 20 countries with the most resilient transport and logistics systems:

- An increase in economic openness to globalization through accelerated export growth of 957.64% (reaching 21.10%);
- A strengthening of national brand marketing power by 29.02% (reaching 8.36 points);
- A rise in the prevalence and integration of digital technologies within companies by 24.75% (reaching 7.73 points).

As a result, the share of transport and logistics services in commercial service exports will reach 68.02%, indicating a 63.64% increase in the resilience of the transport and logistics systems of the studied countries, thus confirming the viability of implementing the infrastructure approach.

5 Discussion

The findings and conclusions of this research enhance the sustainable development concept in transport and logistics systems, contributing to the findings of Abdurakhmanova et al. [2, 3], Amrin et al. [4], Andekina et al. [5], Galoyan and Matevosyan [8], and Mkrtchyan et al. [10]. The core of the scientific novelty of this research is the innovative infrastructure approach developed for state regulation of transport and logistics system development, which is compared with existing literature in Table 2.

According to Table 2, the availability of digital talent (contrary to Popkova et al. [12]), financial support for innovation (contrary to Smetanina and Morozova [13]; Tursunov [16]), diffusion of advanced (5G) telecommunications (contrary to Bogoviz et al. [6]), and social engagement in the Fourth Industrial Revolution (contrary to Sukhodolov et al. [14]) have a restraining effect on the resilience of transport and logistics systems.

Simultaneously, economic openness to globalization and exports (in support of Abaidullayeva et al. [1]; Ergasheva et al. [7]), the marketing power of national brands (in support of Osovtsev et al. [11]), and the prevalence and integration of digital technologies within companies (in support of Sultanova and Babakhanova [15]) positively influence the resilience of transport and logistics systems.

Table 2 Infrastructure approach to transport logistics sustainability similarities and differences from the literature

Structural elements of the transport logistics infrastructure system in the context of the Fourth Industrial Revolution	Coverage of these structural blocks in the literature	Influence of the structural element of infrastructure on the sustainability of transport logistics	
		Essence of influence	Regression coefficient
Availability of digital talents	Popkova et al. [12]	Restraining	-3.84
Financial support for innovation	Smetanina and Morozova [13], Tursunov [16]	Restraining	-3.76
Economic openness to globalization and exports	Abaidullayeva et al. [1] and Ergasheva et al. [7]	Promoting	0.11
Marketing power of national brands	Osovtsev et al. [11]	Promoting	3.86
Prevalence and integration of digital technologies in companies	Sultanova and Babakhanova [15]	Promoting	10.65
Diffusion of advanced (5G) telecommunications	Bogoviz et al. [6]	Restraining	-0.20
Social engagement in the Fourth Industrial Revolution	Sukhodolov et al. [14]	Restraining	-37.19

Source Developed by the author

6 Conclusion

The main conclusion of this research is the critically important role of infrastructure, including digital technologies, in the sustainable development of transport and logistics systems in the context of the Fourth Industrial Revolution. Based on this finding, the author proposed a new infrastructure approach to ensure the sustainable development of transport and logistics systems supported by digital technologies.

The theoretical significance of the proposed approach lies in offering a new set of tools for managing transport logistics, perfectly suited to the Fourth Industrial Revolution: increasing economic openness to globalization, strengthening the marketing power of national brands, and expanding the prevalence and integration of digital technologies in companies.

The practical significance of this approach is that its application will enable countries with less favorable geographic locations to integrate into global trade routes. The proposed approach will also help countries with intergovernmental trade agreements overcome the moral and physical degradation of their transport and logistics systems, thereby enhancing the movement of goods and production factors across their territories.

Drawing on the best practices of the top 20 countries with the most resilient transport and logistics systems, the author's forecast shows that implementing the proposed infrastructure approach will allow countries to achieve sustainable development of transport and logistics systems in the era of the Fourth Industrial Revolution.

References

1. Abaidullayeva M, Khajiyeva G, Karimova M, Arupov A, Zakirova D (2023) Trade and economic 'wars' between the USA and Russia, the consequences for the economy of Kazakhstan. J World Trade 57(4):663–688. https://doi.org/10.54648/TRAD2023026
2. Abdurakhmanova GQ, Astanakulov OT, Goyipnazarov SB, Irmatova AB (2022) Tourism 4.0: Opportunities for applying Industry 4.0 technologies in tourism. In: Proceedings of ICFNDS '22: the 6th international conference on future networks & distributed systems (pp 33–38). New York, NY: Association for Computing Machinery. https://doi.org/10.1145/3584202.3584208
3. Abdurakhmanova GQ, Jeong JY, Oqmullayev RR u, Karimov MU u (2021) The impact of tourism on employment and economic growth in Uzbekistan: An ARDL bounds testing approach. In: Proceedings of ICFNDS '21: The 5th international conference on future networks and distributed systems (pp. 431–439). New York, NY: Association for Computing Machinery. https://doi.org/10.1145/3508072.3508158
4. Amrin AK, Kaliyeva SA, Alzhanova FG (2020) Migration processes in Kazakhstan in the context of the formation of the silk road economic belt. J Popul Soc Stud 28(2):156–174. https://doi.org/10.25133/JPSSV28N2.011
5. Andekina R, Rakhmetova R, Kaliyeva S, Maxyutova A (2024) Entrepreneurial universities of the Republic of Kazakhstan: The essence and features of development. Sci Her Uzhhorod University Ser: Phys, 55, 2690–2700. https://doi.org/10.54919/physics/55.2024.269an0

6. Bogoviz AV, Lobova SV, Ragulina YuV, Luchitskaya LB, Shutova TV (2017) Boosting innovative activity in companies: Problems and potential. J Appl Econ Sci 12(6):1690–1701
7. Ergasheva ST, Tillyakhodjaev AA, Karrieva YK, Popkova EG, Gornostaeva ZV (2023) Business risks in COVID-19 crisis dataset modeling: Regulatory versus marketing tools of risk management. Risks, 11(11), 190. https://doi.org/10.3390/risks11110190
8. Galoyan D, Matevosyan D (2023) The impact of the Russian-Ukrainian conflict on Armenia's foreign trade: A quantitative analysis. J Lib Int Aff 9(2):89–103. https://doi.org/10.47305/JLIA2392106g
9. IMD (2024) World competitiveness online. Retrieved from https://worldcompetitiveness.imd.org/customsearchresults/criteriaresult (Accessed 12 November 2024)
10. Mkrtchyan TM, Tovmasyan GR, Dallakyan SA (2023) Sustainable development of the tourism sector of the Republic of Armenia in the context of an innovative economy. In E.G. Popkova (Ed.), Smart Green Innovations in Industry 4.0 (pp. 383–391). Cham, Switzerland: Springer. https://doi.org/10.1007/978-3-031-45830-9_43
11. Osovtsev VA, Przhedetskaya NV, Sagidullaeva MS (2018) Conceptual model of adaptive management of strategic marketing: A system approach. Eur Res Stud J 21:666–677
12. Popkova EG, Mannapova RA, Troyanskaya MA, Tyurina Yu G (2024) The role of knowledge and technology management in improving the quality of business 4.0 products. Int J Qual Res, 18(1), 167–186. https://doi.org/10.24874/IJQR18.01-11
13. Smetanina AI, Morozova IA (2015) Infrastructural support of electronic payment systems markets in Russia. Appl Econ Int Dev 15(1):57–68
14. Sukhodolov AP, Popkova EG, Litvinova TN (2018) Basic characteristics of information economy. In Sukhodolov AP, Popkova, EG, Litvinova TN (Eds), Model Mod Inf Econ: Concept Contradict Pract Exs (pp. 17–24). Leeds, UK: Emerald Publishing Limited. https://doi.org/10.1108/978-1-78756-287-520181003
15. Sultanova S, Babakhanova N (2023) Statistical model of operational costs indicators in the intelligent transport system of JSC "Uzbekistan railways." AIP Conf Proc 2612(1):060014. https://doi.org/10.1063/5.0119125
16. Tursunov B (2022) Financial security in small business in period of digital economy: In case of Uzbekistan. Proceedings of ICFNDS '22: The 6th international conference on future networks & distributed systems (pp. 491–498). New York, NY: Association for Computing Machinery. https://doi.org/10.1145/3584202.3584275
17. World Integrated Trade Solution (WITS) (2024) Transport services (% of commercial service exports). Retrieved from https://wits.worldbank.org/CountryProfile/en/country/by-country/startyear/LTST/endyear/LTST/indicator/TX-VAL-TRAN-ZS-WT (Accessed 12 November 2024)

Social Communications within Transport Logistics for Sustainable Development Aligned with the ESG Agenda

Development of the Institution of Social Partnership in the Regional Economy of Russia for the Sustainable Development of Transport and Logistics in Support of the Implementation of the ESG Agenda

Ruslan V. Sharifov, Denis E. Matytsin, and Agnessa O. Inshakova

Abstract The research aims to identify prospects for strengthening the resilience of transport and logistics in support of the ESG agenda through the development of the institution of social partnership in the regional economy of Russia. Based on the experience of the Southern and North Caucasus Federal Districts of the Russian Federation in 2023, the authors developed an econometric model to determine the relationship between the E, S, and G outcomes of sustainable transport and logistics development in Russian regions and the effectiveness of key activities of the institution of social partnership. The model revealed patterns in achieving the ESG agenda in the transport and logistics market as the institution of social partnership evolves, underscoring its theoretical significance. The primary conclusion drawn by the authors is that such activities of the institution of social partnership as ensuring employment for male workers and expanding employment opportunities in rural areas contribute to the achievement of the ESG agenda in the transport and logistics market of Russia's regions. Based on the constructed model, the authors determined a prospect for maximizing the implementation of the ESG agenda in Russian regional economies by enhancing the resilience of transport and logistics (in terms of the E, S, and G components) by optimizing the development of the institution of social partnership. The practical significance of this prospect lies in its potential to improve the efficiency of managing sustainable transport and logistics development in the regions of the Southern and North Caucasus Federal Districts of the Russian Federation.

R. V. Sharifov
Volgograd State Technical University, Volgograd, Russia
e-mail: sharifov_rv@mail.ru

D. E. Matytsin · A. O. Inshakova (✉)
Kuban State University, Krasnodar, Russia
e-mail: ainshakova@list.ru

Volzhsky Institute of Economics, Pedagogy and Law, Volzhsky, Russia

D. E. Matytsin
e-mail: dmatytsin@yandex.ru

S. M. Sultanova et al. (eds.), *Sustainable Development of Transport*, Lecture Notes in Networks and Systems 1366, https://doi.org/10.1007/978-3-031-88846-5_19

Keywords Social partnership · Regional economy of Russia · Sustainable development · Transport and logistics · ESG agenda

JEL Codes O18 · Q01 · Q56 · R11 · R13 · R23

1 Introduction

The transport and logistics market is closely interconnected with the labor market, a relationship that is particularly evident at the regional economic level within countries. The importance of studying the transport logistics market lies in the fact that it constitutes a critical area of infrastructure to which regional economies are highly sensitive. Crises in the transport and logistics market can, and often do, have a systemic impact on regional economies, triggering what is commonly referred to as a "domino effect."

Such crises impede the delivery of goods to their destinations, causing shortages of imported goods and overproduction of exported products, which cannot be sold. Disruptions in transport links lead to the paralysis of regional industries, which face shortages of essential production resources, including raw materials, components, supplies, and workers who cannot reach their workplaces.

Despite ongoing automation, organizations within the transport and logistics market continue to be major employers in Russian regions. Given the current level of technological development, opportunities for remote work in this market are significantly limited and occur only in isolated cases. The transport and logistics sector offers employment for workers with various levels of qualifications, including knowledge-intensive, high-productivity, and well-paid positions, as well as jobs for digital specialists.

Thus, Russian regions have a vested interest in the sustainable development of transport and logistics. A pressing scientific and practical issue is the imperfect approach to managing the sustainable development of transport and logistics currently applied in the regional economy of Russia. The existing approach interprets the sustainability of transport and logistics primarily from the perspective of their uninterrupted operation, ensuring stable infrastructural support for the region's economy.

One of the main shortcomings of the current approach is the isolated management of sustainable development in the transport and logistics market, detached from the regional labor market. As a result, personnel management in transport and logistics organizations treats employees as secondary production resources. Without a sense of their value to these organizations, workers exhibit low loyalty toward their employers.

Transport and logistics organizations are not focused on preserving jobs during automation, readily reducing staff, and practicing cost-cutting measures in wages, which diminishes the attractiveness of these positions for regional workers. This approach inevitably leads to a talent drain and an inability of transport and logistics organizations to maintain their competitive edge. The issue is particularly acute in public transportation, which is predominantly state-owned in Russian regions, leading to price caps and further wage suppression.

The adoption of the ESG agenda by modern Russian regions calls for a rethinking of the interpretation and approach to achieving sustainable development in transport and logistics within the regional economy. The new ESG vision for the sustainable development of the transport and logistics market involves a deeper integration with the regional labor market. The institution of social partnership could potentially play a role in facilitating this integration. To contribute to solving this problem, the article aims to define prospects for strengthening the resilience of transport and logistics in support of the ESG agenda through the development of the institution of social partnership in the regional economy of Russia.

2 Literature Review

The theoretical foundation of the research in this article is the scientific concept of ESG development in regional economies [6, 12]. According to this concept, the sustainable development of the transport and logistics market entails organizing the market in line with the region's ESG agenda [8]. In the Russian regional economy, the ESG agenda for the transport and logistics market includes the following:

- E: Improving environmental conditions by enhancing the ecological sustainability of transport and logistics, achieved through more efficient capture and neutralization of air pollutants [11];
- S: Increasing the social responsibility of transport and logistics organizations by creating and maintaining jobs in the region to reduce unemployment among the potential workforce [9];
- G: Enhancing the economic efficiency of corporate governance in organizations operating in the regional transport and logistics market, thereby reducing the share of unprofitable organizations in this market [10].

As a promising tool for integrating the markets for goods and services with the labor market in regional economies, the existing literature highlights the institution of social partnership. In the transport and logistics market, this institution represents a partnership between employers in regional transport and logistics organizations, their workers, and regional public administration bodies that regulate both the transport and logistics market and the labor market [5]. The key activities of the institution of social partnership identified in the literature include the following:

- Ensuring employment for male workers [2];
- Supporting employment for female workers [1];
- Creating the most favorable conditions for urban employment [4];
- Expanding employment opportunities in rural areas [3, 14].

As revealed by the literature review, despite extensive coverage of the issues related to the sustainable development of transport and logistics, as well as the functioning of the institution of social partnership in the regional economy, the contribution of the identified activities of the institution of social partnership to the implementation of the ESG agenda in the transport and logistics market has not been clearly defined and remains uncertain. This gap in the literature prompts the following research question: "Which activities of the institution of social partnership contribute to achieving the ESG agenda in the transport and logistics market in the regions of Russia?".

3 Materials and Methods

The research subjects are the regions within the Southern and North Caucasus Federal Districts of the Russian Federation. The scientific rationale for selecting these regions is based on the following:

1. The sample size is sufficiently large for econometric analysis, comprising 15 regions;
2. The sample includes regions with varying levels of ESG development, ensuring its representativeness and the ability to generalize conclusions and recommendations to the Russian regional economy.

The time frame of the research is 2023 (using data from the end of 2022). Through regression analysis, the study determines the relationship between the success indicators of the ESG agenda implementation in the transport and logistics market (E: the share of captured and neutralized atmospheric pollutants; S: unemployment among the potential workforce; and G: the share of unprofitable organizations) and the success indicators of the main activities of the institution of social partnership (employment of male workers, employment of female workers, urban employment, and rural employment).

Additionally, a regression model is developed to analyze the dependence of each region's overall ESG development level on the aforementioned E, S, and G factors individually. Correlation analysis is used to identify the interconnection between the studied indicators, aiming to establish the interdependence between the transport and logistics market and the labor market in Russian regions. The statistics used in the research are summarized in Table 1.

Based on the results of the analysis, the authors selected the success indicators of the main activities of the institution of social partnership (ScPts) that simultaneously support the improvement of E, S, and G outcomes in the sustainable development of transport and logistics in Russian regions. The prospect of maximizing the implementation of the ESG agenda in Russia's regional economy is determined through enhancing the resilience of transport and logistics (in terms of the E, S, and G components) by optimizing the development of the institution of social partnership.

Table 1 Institute for social partnership, sustainability of transport and logistics, and implementation of ESG agenda in Russian regions in 2023

Federal District	Region	ESG level of the regional economy, score	Employment of male employees, %	Employment of female employees, %	Employment of workers in cities, %	Employment of workers in rural areas, %	Share of captured and neutralized air pollutants, %	Unemployment among potential labor force, %	Share of unprofitable transport and logistics organizations, %
		L_{ESG}	$ScPts_1$	$ScPts_2$	$ScPts_3$	$ScPts_4$	E	S	G
Southern Federal District	Republic of Adygeya	0.467	57.2	43.9	52.7	47.4	40.0	8.5	12.8
	Republic of Kalmykia	0.403	63.7	52.2	56.1	58.8	60.0	9.3	3.3
	Krasnodar Territory	0.662	66.2	50.6	59.5	55.3	29.8	4.6	75.1
	Astrakhan Region	0.477	67.0	51.7	61.1	53.8	41.9	8.4	11.8
	Volgograd Region	0.422	67.0	49.9	59.4	51.6	40.9	4.8	38.1
	Rostov Region	0.541	66.6	52.6	61.5	53.5	40.8	5.0	69.9
North Caucasus Federal District	Republic of Daghestan	0.393	61.6	44.6	54.5	51.0	30.0	18.2	68.8
	Republic of Ingushetia	0.393	57.1	48.1	56.3	46.7	75.0	30.3	0.1
	Kabardino-Balkarian Republic	0.451	65.3	52.7	58.1	58.9	50.0	11.2	3.4
	Karachayevo-Chirkassian Republic	0.454	59.6	45.7	52.5	51.6	50.0	12.1	79.2
	Republic of North Ossetia	0.409	58.0	45.2	50.8	51.3	63.6	15.4	32.1
	Chechen Republic	0.486	58.8	54.2	56.6	56.4	42.9	12.4	1.7
	Stavropol Territory	0.450	64.7	51.4	59.3	54.8	25.0	5.2	44.0

Source Developed by the authors based on the materials from the National Rating Agency [13] and Rosstat [7]

4 Results

To determine the contribution of each activity of the institution of social partnership to achieving the ESG agenda in the transport and logistics market in Russian regions in 2023, the authors conducted a regression analysis of the statistics from Table 1. As a result, the following econometric model was developed:

$$\begin{cases} L_{ESG} = 0.53 + 0.0001 * E - 0.01 * S - 0.0001 * G,\ R^2 = 0.5262; \\ E = -14.48 + 4046 * ScPts_1 - 7.31 * ScPts_2 + 1.44 * ScPts_3 + 0.94 * ScPts_4,\ R^2 = 0.6789; \\ S = 81.86 - 1.82 * ScPts_1 - 0.28 * ScPts_2 + 1.07 * ScPts_3 - 0.06 * ScPts_4,\ R^2 = 0.7463; \\ G = 157.27 - 0.21 * ScPts_1 + 2.14 * ScPts_2 - 3.04 * ScPts_3 - 0.58 * ScPts_4,\ R^2 = 0.6136. \end{cases} \quad (1)$$

Model (1) provides the following conclusions. Collectively, the considered E, S, and G factors account for 52.62% of the overall ESG development level of Russian regions in 2023, meaning that these factors explain a significant portion of this level. An increase in the share of captured and neutralized atmospheric pollutants (E) raises the ESG development level of Russian regions in 2023 by 0.0001. A reduction in unemployment among the potential workforce (S) increases the ESG development level by 0.01. A decrease in the share of unprofitable organizations (G) raises the ESG development level by 0.0001.

A 1% increase in male employment leads to a 4.46% rise in the share of captured and neutralized atmospheric pollutants (E), a 1.82% reduction in unemployment among the potential workforce (S), and a 0.21% decrease in the share of unprofitable organizations (G). A 1% increase in female employment results in a 7.31% decrease in the share of captured and neutralized atmospheric pollutants (E), a 0.28% reduction in unemployment among the potential workforce (S), and a 2.14% increase in the share of unprofitable organizations (G).

A 1% increase in urban employment causes a 1.44% rise in the share of captured and neutralized atmospheric pollutants (E), a 1.07% increase in unemployment among the potential workforce (S), and a 3.04% reduction in the share of unprofitable organizations (G). A 1% increase in rural employment contributes to a 0.94% rise in the share of captured and neutralized atmospheric pollutants (E), a 0.06% reduction in unemployment among the potential workforce (S), and a 0.58% decrease in the share of unprofitable organizations (G).

Thus, the success indicators of the institution of social partnership that simultaneously support the improvement of E, S, and G outcomes in the sustainable development of transport and logistics in Russia's regions are male employment ($ScPts_1$) and rural employment ($ScPts_4$).

Based on Model (1), the prospect of maximizing the implementation of the ESG agenda in Russia's regional economy through enhancing the resilience of transport and logistics (in terms of E, S, and G components) by optimizing the development of the institution of social partnership is outlined in Fig. 1.

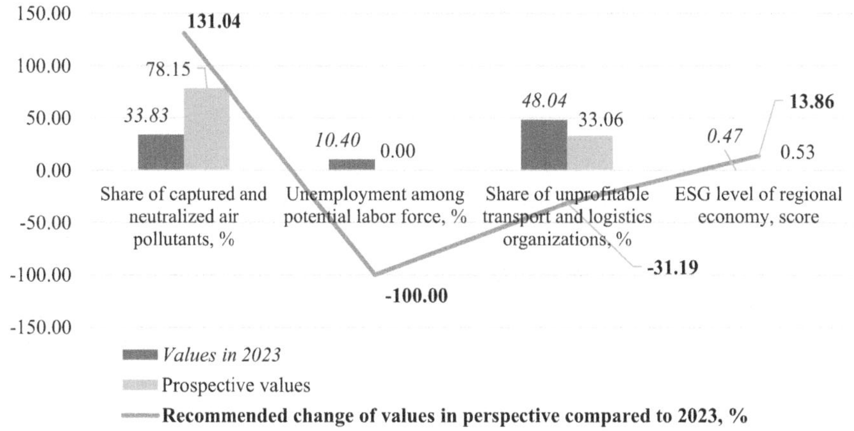

Fig. 1 The prospect of maximizing the implementation of the ESG agenda in the Russian regional economy through enhancing the resilience of transport and logistics. *Source* Developed by the authors

As illustrated in Fig. 1, optimizing the development of the institution of social partnership makes it possible to systemically increase the resilience of transport and logistics in Russian regions, achieving the following:

- An increase in the share of captured and neutralized atmospheric pollutants (E) by 131.04%, from 33.83% in 2023 to 78.15%;
- The complete elimination of unemployment among the potential workforce (S), compared to 10.40% in 2023;
- A reduction in the share of unprofitable organizations (G) by 31.19%, from 48.04% in 2023 to 33.06%.

As a result, in the given outlook, the overall ESG development level of Russian regions in 2023 will rise by 13.86%, from 0.47 to 0.53. The requirements for optimizing the development of the institution of social partnership to enhance the resilience of transport and logistics in support of the ESG agenda in Russia's regional economy are presented in Fig. 2.

Guided by Fig. 2, to optimize the development of the institution of social partnership with the goal of enhancing the resilience of transport and logistics in support of the ESG agenda in Russia's regional economy, the following is recommended:

- Increase male employment by 7.69%, from 63.15% in 2023 to 68.00%;
- Boost rural employment by 48.57%, from 49.71% in 2023 to 73.86%.

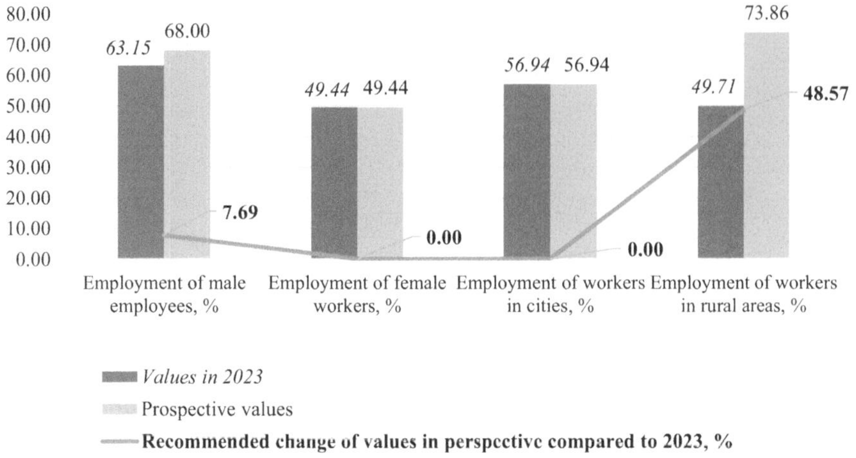

Fig. 2 Requirements for optimizing the development of the institution of social partnership to enhance the resilience of transport and logistics in support of the ESG agenda in Russia's regional economy. *Source* Developed by the authors

5 Discussion

The research contributes to the works of Ergasheva et al. [6] and Samieva et al. [12] on the scientific concept of ESG development in regional economies by clarifying and specifying the contribution of various activities of the institution of social partnership to the implementation of the ESG agenda in the transport and logistics market within the regional economy of Russia. The obtained results are compared with the existing literature in Table 2.

As noted in Table 2, contrary to Abdumalikova [1], it was found that supporting female employment has a contradictory impact on the implementation of the ESG agenda in the transport and logistics market in Russian regions, reducing unemployment among the potential workforce but decreasing the share of captured and neutralized atmospheric pollutants and increasing the share of unprofitable transport and logistics organizations.

In contrast to the findings of Bogoviz et al. [4], it was revealed that creating the most favorable conditions for urban employment has a mixed effect on the realization of the ESG agenda in the transport and logistics market in Russia's regions, increasing the share of captured and neutralized atmospheric pollutants and reducing the share of unprofitable transport and logistics organizations, but also raising unemployment among the potential workforce.

Supporting the conclusions of Abdurakhmanova et al. [2], this research substantiates that ensuring male employment consistently supports the achievement of the ESG agenda in the transport and logistics market in Russian regions. In line with the research of Bannikova et al. [3] and Tursunov [14], it has been proven that expanding employment opportunities in rural areas systemically supports the achievement of

Table 2 Results obtained in contrast to the literature

Areas of activity of the institute of social partnership in the regional economy	Literature suggesting the positive impact of these areas on the ESG agenda of the transportation and logistics market in the regional economy	Actual impact of these areas on the ESG agenda of the transportation and logistics market in Russian regions (regression coefficient)		
		On E: capture and neutralization of air pollutants	On S: unemployment in the potential labor force	On G: the share of unprofitable transport and logistics organizations
Ensuring employment for male workers	Abdurakhmanova et al. [2]	Positive (4.46)	Positive (-1.82)	Positive (−0.21)
Supporting employment for female workers	Abdumalikova [1]	Negative (-7.31)	Positive (-0.28)	Negative (2.14)
Creating the most favorable conditions for urban employment	Bogoviz et al. [4]	Positive (1.44)	Negative (1.07)	Positive (−3.04)
Expanding employment opportunities in rural areas	Bannikova et al. [3] and Tursunov [14]	Positive (0.94)	Positive (-0.06)	Positive (−0.58)

Source Developed by the authors

the ESG agenda in the transport and logistics market of the modern Russian regional economy. Thus, the research contributes to the scientific discussion led by Inshakova et al. [8], Jalolova et al. [9], Matytsin et al. [10], and Menglikulov et al. [11] on the sustainable development of the transport and logistics market in accordance with the ESG agenda of the region.

6 Conclusion

Thus, the achievement of the ESG agenda in the transport and logistics market in Russian regions is supported by the activities of the institution of social partnership, such as ensuring male employment and expanding employment opportunities in rural areas.

The theoretical significance of this research lies in the developed econometric model, which establishes the relationship between the E, S, and G outcomes of sustainable transport and logistics development in Russian regions and the success of implementing the key activities of the institution of social partnership. This model reveals the patterns of achieving the ESG agenda in the transport and logistics market as the institution of social partnership evolves.

The identified prospect of maximizing the implementation of the ESG agenda in Russia's regional economy by enhancing the resilience of transport and logistics (in terms of the E, S, and G components) through optimizing the development of the institution of social partnership holds practical significance for the regions of the Southern and North Caucasus Federal Districts of the Russian Federation because it will improve the efficiency of managing sustainable transport and logistics development in these regions.

References

1. Abdumalikova G-F (2022) Demographic changes affecting the female dividend in the digital economy. In: Proceedings of ICFNDS '22: The 6th international conference on future networks & distributed systems (pp. 581–584). New York, NY: Association for Computing Machinery. https://doi.org/10.1145/3584202.3584289
2. Abdurakhmanova GQ, Jeong JY, Oqmullayev RR, Karimov MU (2021) The impact of tourism on employment and economic growth in Uzbekistan: An ARDL bounds testing approach. In: Proceedings of ICFNDS '21: The 5th international conference on future networks and distributed systems (pp. 431–439). New York, NY: Association for Computing Machinery. https://doi.org/10.1145/3508072.3508158
3. Bannikova NV, Astrakhantseva EY, Morozova IA, Litvinova TN (2019) Mastering of the information and communication technologies by labor resources of rural territories as a new vector of their development in modern Russia. In Popkova E, Ostrovskaya V (Eds). Perspectives on the use of new information and communication technology (ICT) in the modern economy (pp. 701–708). Cham, Switzerland: Springer. https://doi.org/10.1007/978-3-319-90835-9_81
4. Bogoviz AV, Ragulina Yu V, Alekseev AN, Lavrov MN, Kletskova EV (2018) Economic stimuli for creation of highly-efficient jobs for a modern human. In Popkova E (Ed), Impact Inf Mod Hums (pp. 624–630). Cham, Switzerland: Springer. https://doi.org/10.1007/978-3-319-75383-6_80
5. Dzhumanova AB, Kushakova MN, Khodzhaeva NA (2019) Formation of accounting management information in the control system of enterprises of JSC "Uzbekistan Railways." Int J Adv Sci Technol 28(14):32–36
6. Ergasheva ST, Khaydarova NA, Litvinova TN, Cherepukhin T Yu (2023) ESG management of robotized business in support of green economy development in Russia and Central Asia.) In Popkova EG, Sergi BS (Eds). ESG management of the development of the green economy in Central Asia (pp. 13–21). Cham, Switzerland: Springer. https://doi.org/10.1007/978-3-031-46525-3_2
7. Federal State Statistics Service of the Russian Federation (Rosstat) (2023) Regions of Russia. Socio-economic indicators 2023. Moscow, Russia: Rosstat. Retrieved from https://rosstat.gov.ru/storage/mediabank/Reg_Rus_Pokaz_2023.htm (Accessed 26 August 2024)
8. Inshakova AO, Matytsin DE, Inshakova EI (2024) State regulation policy in the investment sphere as a key factor in ensuring inclusive growth and sustainable economic development. In Inshakova A, Matytsin D, Inshakova E (Eds), Remote investment transactions in the digital age (pp 3–9). Cham, Switzerland: Springer. https://doi.org/10.1007/978-3-031-51536-1_1
9. Jalolova M, Sangirova U, Yakubov I, Rahimov H, Kholmatova N (2022) Economic efficiency of the transport system and logistics in the Republic of Uzbekistan. Transp Res Procedia 63:1061–1066. https://doi.org/10.1016/j.trpro.2022.06.107
10. Matytsin DE, Inshakova AO, Goncharov AI (2023) Transformation of settlement relations in the context of industry 4.0: Conversion of blockchain club's crypto-codes into legitimate analogues.

In Popkova EG (Ed), Sustain Dev Risks Risk Manag (pp. 43–46). Cham, Switzerland: Springer. https://doi.org/10.1007/978-3-031-34256-1_8

11. Menglikulov B, Umarov S, Safarov A, Zhyemuratov T, Alieva N, Durmanov A (2023) Ways to increase the efficiency of transport logistics—Communication services in Uzbekistan. E3S web of conferences, 389, 05036. https://doi.org/10.1051/e3sconf/202338905036
12. Samieva KT, Saenko II, Menshchikova VI, Smetanin AS (2023) ESG Management of digital business using big data and artificial intelligence (AI) in support of the green economy in Russia and Central Asia. In Popkova EG, Sergi BS (Eds.), ESG management of the development of the green economy in Central Asia (pp. 3–11). Cham, Switzerland: Springer. https://doi.org/10.1007/978-3-031-46525-3_1
13. National Rating Agency (2022) Ranking of sustainability of development of the Russian Federation and integration of ESG-criteria in the activities of the subjects of the Russian Federation: Analytical review. Retrieved from https://www.ra-national.ru/wp-content/uploads/2022/12/ranking_esg_regions.pdf (Accessed 26 August 2024)
14. Tursunov BO (2022) Provincial features of industrial production dynamics in the research of textile enterprises' financial security in Uzbekistan. In Popkova EG (Eds). Imitation market modeling in digital economy: Game theoretic approaches (pp 601–610). Cham, Switzerland: Springer. https://doi.org/10.1007/978-3-030-93244-2_66

Methodology for Assessing the Marketing Potential of "O'ztemiryo'lyo'lovchi" JSC

Ziyodullo N. Raxmatov and Sitora Sh. Raxmatova

Abstract Recent trends evidence the growing competition between modes of transport in the Republic of Uzbekistan, driven by the development of the railway sector. In the context of growing competition, the management of "O'ztemiryo'lyo'lovchi" JSC (Tashkent, Uzbekistan) needs to introduce the most advanced approaches and methods of organizing passenger transportation. This research discusses a systemic analysis of existing approaches to assessing the marketing potential of the methodology for assessing the marketing potential of the developer, reflecting the potential of the company's marketing communications, human resources, market potential, and financial potential. The authors highlight the role of marketing activities in increasing the competitiveness of the rail transport sector. The improvement in transport activities is associated with the development of the company's marketing and management structures. Assessing the marketing potential of railway transport makes it possible to study the impact of marketing communications on its commercial activities. This is a key condition for improving them and increasing the level of impact on consumers. Thus, the research aims to determine what factors affect the company's marketing potential. Additionally, the research examines advanced marketing strategies to promote the railway sector. The analyzed factors also make it possible to determine the possibilities of future marketing activities and choose the direction of the company's development.

Keywords Marketing potential · Assessment · Marketing communications potential · Human resource · Financial potential

JEL codes M30 · M31 · M39 · L91 · L92

Z. N. Raxmatov (✉) · S. Sh. Raxmatova
Tashkent State Transport University, Tashkent, Uzbekistan
e-mail: znraxmatov@tstu.uz

S. Sh. Raxmatova
e-mail: sitmeh18@gmail.com

S. M. Sultanova et al. (eds.), *Sustainable Development of Transport*, Lecture Notes in Networks and Systems 1366, https://doi.org/10.1007/978-3-031-88846-5_20

1 Introduction

To thrive in a competitive market, economic systems must shift from merely increasing resources and assets to embracing innovative and technological advancements. This transition requires a thorough and accurate analysis of key factors that influence economic growth. Among these factors, marketing plays a crucial role, focusing on understanding consumer behavior and building competitive advantages in the market. A key measure of economic development is its potential, which is largely shaped by one of its most significant components—marketing potential [9].

In recent research, identifying the transport component within enterprise marketing potential has become a critical task. The transport aspect of marketing potential, along with its various components, helps determine how transport enterprises will evolve and highlights areas for improving the effectiveness of marketing strategies. However, the lack of a unified formula or methodology developed by both foreign and local scholars for assessing marketing potential makes this task even more relevant and challenging.

When carrying out marketing activities, transport enterprises in competitive market conditions need a marketing event management system developed based on scientific recommendations for evaluating and analyzing the event. For this purpose, it is necessary to scrupulously study the marketing potential of a transport enterprise.

E. N. Kolisnik, through a systemic analysis of current research on evaluating marketing potential, proposed a model to enhance marketing capacity. This model encompasses internal and external factors of the enterprise that influence its marketing potential, including available resources, market opportunities, and competitive conditions. By assessing the future marketing potential of an enterprise, the model enables the selection of optimal marketing strategies and the effective use of marketing opportunities for growth and development [6].

In the scientific literature, various approaches have been proposed to assess the marketing potential of transport enterprises. A. Rojkov, in his research, suggested using indicators such as market size, sales volume, advertising expenses, and competition costs [10]. It was proven that these factors make up the marketing potential.

The increasing use of personal cars among the population leads to problems such as traffic jams on city streets and air pollution. The rise in population income, combined with the widespread belief that cars are the most convenient mode of transport, has led to increased competition for companies offering passenger transport services. In this regard, it is essential to develop marketing activities in public transport, maximize the use of its tools, and reduce the use of personal transport. The studies conducted by A. Ibraeva and J. F. Sosa proposed additional functions to enhance the practical application of the economic concept of "marketing potential" in transport organizations. These additions aim to better align marketing potential with the specific needs and dynamics of the transport sector. To determine the marketing potential, it is recommended to use the level of marketing security, the potential of marketing tools, the potential of marketing department employees, and the financial potential of marketing activities [4].

Local scholars such as [1–3, 5, 7, 11] have proposed various methods related to transport services, marketing strategies, innovation marketing, and marketing communications. However, while their work provides valuable insights, it does not fully explore the marketing potential specific to the railway passenger sector. Additionally, the methodology for assessing this potential has not been thoroughly investigated.

2 Materials and Methods

To assess the marketing potential of "O'ztemiryo'lyo'lovchi" JSC (Tashkent, Uzbekistan), a methodology is proposed based on the analysis of foreign and local scientific research. This approach evaluates all factors influencing the company's marketing potential, considering key elements such as the potential of marketing communications (I_{mpc}), human resources potential (I_{sp}), market share potential (I_{mpc}), and financial potential ($I_{,fn}$). The company's overall marketing potential will be visualized using a geometric mean, allowing for the analysis of individual indicators measured in different units. This method provides a clear view of the dynamics of each factor and the overall marketing potential.

3 Results

To assess the marketing potential (I_{mpc}), the authors propose indicators that characterize the company's marketing activities, the knowledge gained, and the implementation of new ideas. These indicators are as follows:

(a) An indicator of the level of organization of marketing communication activities. In contemporary marketing theory and practice, the importance of marketing communications is gradually increasing; companies and firms in the competitive market realize that the importance of communications in solving their specific problems is high. Marketing communications play an important role in forming and maintaining relations of goods and services of companies with real and potential consumers in a free economy [9]. To determine the level of the organization of the company's marketing communication activities, it is possible to analyze the distribution of net profit over a given period. This analysis involves comparing the net profit generated with the costs incurred from marketing communication efforts during that same timeframe. By examining this relationship, it becomes possible to assess how efficiently the company has managed and executed its marketing communication activities in relation to its financial outcomes. It is calculated using the following formula:

$$K_{lmp} = \frac{C_{tmc}}{A_{np}}, \tag{1}$$

where:
C_{tmc}—total marketing costs;
A_{np}—net profit.

(b) The level of response to marketing events. There are also opinions that the effectiveness of marketing communications is directly related to the cost of sales of goods and services [8]. This investment does not always make sense because advertising and marketing are other elements of communication that customers may simply like and do not affect the idea of purchasing their goods and services [6]. It is determined by the ratio of the number of marketing communication events offered by the employees of the company's marketing department to the number of events taken by management. This indicator is calculated using the following formula (2):

$$K_{lmas} = \frac{Q_{nmr}}{Q_{nm}} \tag{2}$$

where:
Q_{nmr}—the number of measures taken in marketing communications;
Q_{nm}—the number of proposed activities in the field of marketing communications.

(c) An indicator of the acceptance of marketing communication activities by passengers. Marketing conducted by the company among passengers determines the response to communication activities. It is calculated according to the following formula (3):

$$K_{pam} = \frac{Q_{pum}}{The_{npc}}, \tag{3}$$

where:
Q_{pum}—the number of passengers using marketing communication activities;
The_{npc}—total number of passengers transported.

The marketing potential index reflects the integral indicator of its individual indicators.

$$I_{mpc} = \sqrt[3]{K_{lmp} + K_{lmas} + K_{pam}}, \tag{4}$$

The next indicator is human resources potential (I_{sp}). The authors suggest using descriptive indicators to evaluate the potential of employees, focusing on factors such as the composition of the marketing department staff, administrative personnel, and the proportion of marketing staff involved in periodic training. These indicators include the following:

(a) Marketing communication activities in the development of human resources potential. The share of employees engaged in the development and implementation of marketing communications activities in the company is calculated as follows (5):

$$K_{dpm} = \frac{Q_{nd}}{Q_{nas}}, \quad (5)$$

where:
Q_{nd}—the number of employees of the marketing department;
Q_{nas}—the number of administrative staff.

(b) Indicator of the proportion of employees with a marketing specialty. This indicator determines the number of employees with special knowledge. It is calculated as follows:

$$K_{pms} = \frac{Q_{nsseems}}{The_{nas}}, \quad (6)$$

where:
$Q_{ns\ seems}$—the number of employees specializing in marketing;
The_{nas}—the number of administrative staff.

(c) The indicator of the share of qualified employees of the marketing department. This indicator determines the number of employees who are sent for advanced training from the marketing department during the year. It is calculated as follows:

$$K_{pt} = \frac{Q_{nts}}{Q_{nd}}, \quad (7)$$

where:
Q_{nts}—the number of qualified marketing department employees;
Q_{nd}—the number of marketing department employees.

The index reflects the specific personnel potential through a set of integrated indicators (8):

$$I_{sp} = \sqrt[3]{K_{dnp} + K_{pms} + K)}_{pt}, \quad (8)$$

The next aspect is market share potential (I_{spc}). To assess the company's market share, the authors suggest using indicators that characterize the company's competitiveness in the market. These indicators include the following:

(a) Market share index. The authors determine the number of passengers transported by the company by calculating its share of the total number of passengers in the passenger transportation services market:

$$K_{.msc} = \frac{R_{pt}}{R_{apt}}, \tag{9}$$

where:
R_{pt}—the turnover of passengers of "O'ztemiryo'lyo'lovchi" JSC;
R_{apt}—passenger turnover on all transport types.

(b) The index of the movement of passengers in vehicles. Passenger traffic is determined by the number of inhabitants and the population. This indicator is determined by the ratio of the number of passengers in the country. It is calculated by the following formula:

$$K_{tmp} = \frac{M_{np}}{M_{sp}}, \tag{10}$$

where:
M_{np}—the number of passengers sent by "O'ztemiryo'lyo'lovchi" JSC;
M_{sp}—the country's population.

Thus, the market share potential (I_{spc}) is determined using the following formula:

$$I_{spc} = \sqrt[2]{K_{mcs} + K_{,tmp}}, \tag{11}$$

The next aspect is financial potential ($I_{,fn}$). To assess the company's financial potential, The authors encourage the marketing department staff to utilize expenses related to employee qualifications, total passenger transportation costs, marketing activities, and route adjustments to enhance the quality of services provided to passengers. Additionally, the authors suggest using the costs associated with passenger re-registration to improve employee skills, as well as the total transportation and marketing costs, to ensure high-quality service for passengers. These indicators include the following:

(a) The index of the costs of marketing communication activities. It is important to invest in attracting consumers and maintaining their relevance. It is worth noting that the provision of new types of services generates high costs. Thus, effective advertising of existing services often becomes an attractive alternative. The indicator is calculated as follows (12):

$$K_{pm} = \frac{C_{cm}}{C_{sc}}, \tag{12}$$

where:
C_{cm}—event marketing costs;
C_{sc}—sale costs.

(b) Marketing costs and staff training index. The need for continuous training of personnel arises for several reasons, including the following:

- The rapid development of science and technology;
- Dynamic changes in the external environment;
- The emergence of new requirements for consumers;
- The emergence of new products and services from competitors.

All of the personnel development efforts mentioned above require certain investments, which, over time, lead to increased labor productivity within the company, as well as the generation of new ideas and proposals:

$$K_{l.blast} = \frac{Q_{mstc}}{C_{sc}}, \tag{13}$$

where:

Q_{mstc}—marketing staff training costs;

C_{sc}—sale costs.

The integral index of a company's marketing potential (I_{imc}) is calculated as the geometric mean of indicators of the potential of marketing communications:

$$I_{fn} = \sqrt[2]{K_{pm} + K_{blast,l.}}, \tag{14}$$

The company's marketing potential of the integrated index (I_{imc}) is calculated as the average value of the geometric potential marketing communications indicators (15):

$$I_{imc} = \sqrt{I_{mpc} + I_{sp} + I_{spc} + I_{fn}}. \tag{15}$$

The marketing potential of the proposed method of calculation consists of the following stages:

1. Collecting data and systematizing them;
2. Identifying potential indicators for each element of the integrated market;
3. Calculating the potential of marketing communications.

4 Conclusion

Summarizing the proposed approaches, the following can be concluded:

1. Many scholars agree that evaluating only the economic aspects of efficiency is insufficient [5, 11], indicating that performance appraisal is a multidimensional and complex system. This necessitates incorporating social and even psychological factors when assessing the effectiveness of marketing programs;
2. The concept of evaluating the effectiveness of marketing programs in science with a new cost approach is developing. The result of marketing programs is the creation of intangible marketing tools with commodity capital and interactions (customer capital). Simultaneously, customer capital is a new concept in

marketing programs that have not been considered by scientists and requires conceptual methodological development;
3. To evaluate the effectiveness of marketing programs developed by the company in attracting customers, it will be necessary to develop a single measure that measures marketing effectiveness.

There is no absolute and universal method of evaluating the effectiveness of marketing communications. Therefore, it is important to carefully combine the methods of quality and quantity, considering the influence of internal and external factors.

Thus, assessing marketing potential determines the possible opportunities for further development of "O'ztemiryo'lyo'lovchi" JSC (Tashkent, Uzbekistan). Incorporating factors such as marketing communication potential, human resources, market share potential, and financial potential into the methodology for assessing marketing potential enables a more precise analysis of external and internal marketing factors, as well as the identification of future marketing opportunities.

References

1. Akbarov MI (2011) Management of commercial activities of SIGNED "O'ztemiryo'lyo'lovchi" through marketing communications (Synopsis of Dissertation of Candidate of Economic Sciences). St. Petersburg, Russia
2. Akbarov MI (2010) Justification of the choice of a complex of marketing communications in railway transport on the example of passenger transportation. Proceedings of petersburg transport university 3:175–183
3. Akramov TA (2018) In the car industry and the improvement of marketing strategies (Synopsis of Dissertation of Doctor of Economic Sciences). Tashkent State Univ Econ, Tashkent, Uzbekistan
4. Ibraeva A, de Sauce JF (2014) Marketing of public transport and public transport information provision. Procedia – Soc Behav Sci, 162, 121–128. https://doi.org/10.1016/j.sbspro.2014.12.192
5. Irisbekova MN (2013) Methodological approach to the formation of the marketing strategy of the enterprise car transport. Mark, Money Credit 12:34–40
6. Kolesnik EN (2010) Factors of marketing potential of economic system. In Tatishev readings: Actual problems of science and practice. Materials of VII International scientific-practical conference. Tolyatti, Russia: Volzhsky University named after V. N. Tatishchev
7. Qaxxorov JA (2016) Automobile transport system of Uzbekistan for innovative improvement of marketing activities (Synopsis of Dissertation of Doctor of Philosophy). Tashkent State University of Economics, Tashkent, Uzbekistan
8. Raxmatov ZN (2019) Criteria for evaluating the effectiveness of marketing communications. Econ Innov Technol 4:45–50
9. Raxmatov ZN (2019) Uztemiryo'lyo'lovchi JSC joint-stock company ways of improving the mechanism of development of a marketing communications strategy. Econ Innov Technol 4:165–170
10. Rojkov AA, Besfamilnaya SV (2002) Organization of monitoring of the system of reproduction of jobs created under local development programs in coal-industrial regions. Ugol' [Russian Coal Journal] 3:20–24
11. Ziyayeva MM (2019) Railway passenger transportation network improvement marketing services. Econ Innov Technol 4:185–195

Navigating Success: Language Skills Prove Crucial in Driving the Future of the Transportation Industry

Shakhnoza I. Sheralieva

Abstract The transportation industry is on the cusp of a significant transformation, driven by advancements in automation, electrification, and connectivity. While technological innovation plays a crucial role, soft skills like effective communication and language proficiency are gaining unprecedented importance in navigating this rapidly evolving landscape. This research explores the growing need for language skills in the transportation industry, highlighting key areas where strong communication abilities are essential and showcasing particular examples through tables and charts. The author argues that language skills are a critical driving force for success in the future of transportation. The combined insights from qualitative interviews and job posting analysis provide a compelling insight that navigating the future of transportation demands not just expertise but multilingual fluency. The qualitative topics unequivocally resonate with the findings, as 85% of the interviewees identified language as crucial for communication and collaboration. This sentiment echoes the widespread recognition, with 90% of the participants acknowledging the growing importance of language skills due to globalization.

Keywords Transportation industry · Marketing and public relations · International collaboration · Intercultural communication · Global competitiveness

JEL code O33

1 Introduction

The global transportation industry, valued at a staggering $9.2 trillion in 2023 [1], stands on the precipice of a transformative era. Autonomous vehicles, hyperloop systems, and electric mobility solutions are rapidly transitioning from futuristic concepts to tangible realities, fundamentally altering how people and goods move across the globe [2]. Amidst this technological revolution, a seemingly mundane skill

S. I. Sheralieva (✉)
Tashkent State Transport University, Tashkent, Uzbekistan
e-mail: Shaxa2108@mail.ru

S. M. Sultanova et al. (eds.), *Sustainable Development of Transport*, Lecture Notes in Networks and Systems 1366, https://doi.org/10.1007/978-3-031-88846-5_21

is poised to play a crucial role in navigating the future of transportation: language proficiency.

While the literature on the future of transportation is vast, focusing primarily on technological advancements and economic impacts [3], the importance of soft skills like communication and language proficiency has received less attention. This gap necessitates a closer examination of the role language skills will play in the evolving transportation landscape.

Several studies highlight the growing need for effective communication in the transportation industry. A 2023 World Economic Forum report predicts that 95% of new jobs created by 2025 will require some level of soft skills, including communication and collaboration [3]. Similarly, a 2022 LinkedIn survey found that 73% of hiring managers consider language a valuable skill for new hires [4]. These findings suggest that language skills are not just desirable add-ons but rather essential qualifications for future transportation professionals.

The specific reasons for this growing demand are multifaceted. The globalization of the transportation industry necessitates seamless collaboration across borders and cultures. Study by the International Labour Organization [5] highlights the increasing demand for multilingual professionals in international trade and logistics, sectors intricately linked to transportation. Additionally, the growth of autonomous vehicles and connected transportation systems requires clear and concise communication, both between humans and machines and among the various machine systems [2].

Furthermore, the evolving nature of passenger interaction demands strong communication skills [6]. As automation increases, human contact remains crucial in customer service and passenger-facing roles. Study by the International Air Transport Association [7] emphasizes the importance of cultural sensitivity and effective communication in ensuring positive passenger experiences. While technological advancements drive the future of transportation, the human element remains central. The ability to communicate effectively across cultures and languages will be an essential skill for professionals navigating this evolving landscape [6, 8]. By integrating language skills into workforce development programs and prioritizing multilingual talent acquisition, the transportation industry can ensure its success in the years to come.

2 Methods

This research utilizes a multifaceted approach to explore the rising importance of language skills in the future of the transportation industry. The methods employed include the following:

1. Quantitative analysis:

    - Secondary data analysis: Existing data on the projected demand for multilingual professionals in various transportation segments. The author gathered and analyzed 20 recent industry reports (published in 2020–2024) from the

World Economic Forum, PWC, McKinsey and Company, ITF, and World Bank, as well as academic studies published in journals like Transportation Research and the International Journal of Logistics Management.
- Data extraction: Extracted data on projected demand for multilingual professionals in various segments (autonomous vehicles, logistics, aviation, and public transport) and estimated growth over the next 5–10 years.
- Online job market analysis: The author examined job postings on industry-specific platforms and general job boards like LinkedIn and Indeed. Additionally, the author analyzed the frequency and distribution of keywords related to language skills across different transportation job categories.

2. Qualitative interviews:

The author analyzed in-depth interviews of 20 diverse transportation professionals from a range of sectors and roles, including the following:

- Engineers: Involved in international projects requiring clear communication across languages.
- Project managers: Responsible for leading multicultural teams and navigating cross-cultural challenges.
- Customer service representatives: Interacting with passengers from diverse backgrounds and resolving language-related issues.
- Marketing executives: Developing global marketing campaigns and navigating international partnerships.

Interviews were conducted in March 2023, to capture current perspectives and experiences.

The author utilized semi-structured interviews with open-ended questions to explore the following aspects:

- Personal experiences: Specific challenges and opportunities encountered due to language barriers or multilingual skills;
- Views on importance: Perceptions of the growing need for multilingualism in the industry and its impact on their careers;
- Solutions and suggestions: Ideas for promoting language skills development within the industry and fostering a more inclusive work environment.

By delving into individual experiences and perspectives, the author aimed to gain a nuanced understanding of the following aspects:

- Human impact: How language skills affect daily work, communication, and career advancement in the transportation industry;
- Real-world challenges: Practical difficulties faced due to language barriers and the value of overcoming them;
- Individual perspectives: Opinions and suggestions from those directly involved in the industry, enrich the research with diverse viewpoints.

3 Results and Discussion

First, the author assessed projected growth percentages and key reasons for growth in the demand for multilingual professionals in different transportation segments from 2023 to 2030 (Table 1).

Table 1 provided the following results:

1. The segment of autonomous vehicles is expected to experience a projected growth of 45%. The key reasons for this growth include global development efforts, international collaborations, and the need for technical communication across languages.
2. Logistics and freight are projected to grow by 38%. The expansion of international trade and the increasing complexity of supply chains are the key reasons behind this growth.
3. The demand for multilingual professionals in the public transportation sector is expected to grow by 32%. This growth is driven by increasing ridership from diverse backgrounds, where multilingual customer service plays a vital role.
4. The aviation sector is projected to experience a growth of 28% in the demand for multilingual professionals. This growth is attributed to the increasing trend of international air travel and the necessity for pilots and cabin crew who are proficient in multiple languages.

Table 2 summarizes the topics of interviews related to language skills. The column "Sample Quote" provides a representative quote from one of the interview participants that exemplifies the discussed topic. The "Frequency" column represents the percentage of interview participants who mentioned a particular theme. For example, 85% of the interviewees highlighted the importance of language skills for communication and collaboration, 90% mentioned the growing demand for language skills

Table 1 Projected growth in demand for multilingual professionals by transportation segment (2023–2030)

Transportation segment	Projected growth (%)	Key reasons for growth
Autonomous vehicles	45	Global development, international collaborations, and technical communication across languages
Logistics and freight	38	Expansion of international trade, complex supply chains requiring communication with diverse partners
Public transportation	32	Increasing ridership from diverse backgrounds and multilingual customer service crucial
Aviation	28	Growth in international air travel, need for pilots and cabin crew proficient in multiple languages

Source Developed by the author

due to globalization, 80% referred to challenges stemming from language barriers, 75% highlighted the benefits of multilingualism, and 65% emphasized the need for language training.

The topics of the interview related to language skills in the transportation industry are accompanied by sample quotes from the interviews and the frequency at which each topic was mentioned. Table 2 provides insights into the perspectives and experiences of transportation professionals regarding language skills.

To calculate the percentages in Table 2, the author determined the proportion of a particular value relative to the total or reference value. The general process for calculating percentages is as follows:

1. To determine the total or reference value, the author identified the total reference value used as the basis for calculating the percentages. In Table 2, it is the total number of responses;
2. To calculate the proportion for each category or topic, the author calculated the proportion relative to the total reference value. This is usually done by dividing the specific count or frequency by the reference value;
3. In converting to percentage, the author multiplied the proportion obtained in step 2 by 100. It gave the author the percentage value for each category.

For example, the following is obtained:

- Total or reference value (e.g., total number of interview participants): 20;
- Frequency for "Importance for communication and collaboration": 17 (from the sample quote, assuming it is the count of interviewees mentioning this topic);
- Proportion: 17/20 = 0.85;
- Percentage: 0.85*100 = 85%.

Following this process, the percentages for each category of Table 2 were calculated.

Table 2 Topics of qualitative interview related to language skills

Topic	Sample quote	Frequency (%)
Importance of communication and collaboration	"Essential for clear understanding in international projects" (Engineer)	85
Growing demand due to globalization	"Language skills are becoming a must-have in my field" (Marketing executive)	90
Challenges with language barriers	"Miscommunication leads to delays and misunderstandings" (Project manager)	80
Benefits of multilingualism	"My language skills opened doors to career advancement" (Customer service representative)	75
Need for language training	"Companies should invest in language learning programs"	65

Source Developed by the author

4 Conclusion

The combined insights from qualitative interviews and job posting analysis reveal that navigating the future of transportation demands more than just technical expertise—it requires multilingual fluency.

Qualitative topics align strongly with these findings, as 85% of interviewes highlighted language as essential for effective communication and collaboration. This sentiment is widely shared, with 90% of participants recognizing the increasing importance of language skills in a globalized industry. These findings indicate that multilingualism is no longer merely a bonus but a core competency in transportation.

Language barriers continue to pose challenges, with 80% of respondents reporting delays and misunderstandings due to miscommunication. This highlights the need for proactive measures, as evidenced by 65% of participants advocating for expanded language training programs within companies. Embracing multilingualism thus represents not only a challenge to address but also a significant opportunity to enhance efficiency and foster better understanding.

While English remains a baseline (40%), significant growth is observed in languages like Mandarin (+25%) and Spanish (+12%), mirroring the industry's expanding global reach. French, German, and Arabic also see rising demands, showcasing the need for a diverse language skillset.

Thus, the transportation industry stands at a critical juncture. By actively embracing multilingualism, it can unlock a future of:

- Seamless communication: Clear understanding provides efficient collaboration and reduces costly miscommunications;
- Globalized reach: Catering to diverse customers and navigating international partnerships opens new markets and opportunities;
- Talent pool expansion: A wider language proficiency attracts a broader talent pool, fostering innovation and diverse perspectives;

To embrace the multilingual future, it is imperative for the transportation industry to invest in language training programs, create inclusive work environments, and acknowledge the value of diverse skills [9]. These steps are crucial for ensuring success in an industry that is constantly evolving. It is no longer an option for the industry to ignore this linguistic evolution. Rather, it is a necessity for staying competitive and thriving in the changing landscape of transportation.

References

1. Dickens M (Ed) (2023) 2022 Public transportation fact book (73rd ed) Washington, DC: American public transportation association. https://www.apta.com/wp-content/uploads/APTA-2022-Public-Transportation-Fact-Book.pdf. Accessed 8 Feb 2024

2. Heineke K, Laverty N, Möller T, Ziegler F (2023) The future of mobility. McKinsey and company. https://www.mckinsey.com/industries/automotive-and-assembly/our-insights/the-future-of-mobility-mobility-evolves. Accessed 8 Feb 2024
3. Faggian A, Marzucchi A, Montresor S (2024) Regions facing the twin transition: combining regional green and digital innovations. Reg Stud. https://doi.org/10.1080/00343404.2024.2398555
4. LinkedIn (2022) Global talent trends: the reinvention of company culture. https://business.linkedin.com/content/dam/me/business/en-us/talent-solutions-lodestone/body/pdf/global_talent_trends_2022.pdf. Accessed 8 Feb 2024
5. International labour organization (ILO) (2017) World employment and social outlook: trends for women 2017. https://www.ilo.org/global/research/global-reports/weso/trends-for-women2017/lang--en/index.htm. Accessed 8 Feb 2024
6. Mustaeva G, Ataeva G (2023) The role of business english in modern logistics. E3S Web Conf 402:08008. https://doi.org/10.1051/e3sconf/202340208008
7. International air transport association (IATA) (2023) Global passenger survey (GPS). https://www.iata.org/en/publications/manuals/global-passenger-survey/. Accessed 8 Feb 2024
8. Sheralieva SI, Saydalieva DZ (2021) Application and importance of transport terminology. Turk Online J Qual Inq (TOJQI) 12(10):4742–4750. https://www.tojqi.net/index.php/journal/article/view/8551/6067. Accessed 8 Feb 2024
9. Komkova AS (2021) English language skills university graduates need for employment in transport and logistics industry. Int J Humties Nat Sci 9–2(60):39–41. https://doi.org/10.24412/2500-1000-2021-9-2-39-41
10. Jensen C, Hazelton JK, Wellman G (2020) Finding improvement in the language transportation planners use: a critical discourse analysis to illustrate an automobile-centric bias in transportation policymaking. Public Works Manag Policy 25(2):167–188. https://doi.org/10.1177/1087724X19885937

The Use of Digital Educational Resources in Language Training for Transport Specialists

Svetlana B. Im and Georgiy A. Jakupov

Abstract The education system is one of the most important institutions of today's society because it determines the development of a country. Higher education is considered a leading factor in social, political, and economic progress. In this regard, the work examines ways of modernizing the educational process by digitalization and improving skills in the use of information and communication technologies. Digital educational resources have established themselves as effective tools accompanying traditional learning. The research aims to demonstrate the interactive capabilities of the digital resource center as an auxiliary material in conditions of difficult interaction between the teacher and students. The work used a computer modeling method that makes it possible to reproduce elements of traditional learning using interactive digital learning tools. Moreover, the authors applied diagnostic methods as criteria for assessing the adequacy of means of monitoring the progress of an educational application. The research revealed that the pandemic created an urgent need for computers and other remote communication tools. Under these circumstances, the importance of digital communication centers significantly increased, shifting them from auxiliary tools to the primary means of interaction between students and their study materials. Digital learning centers facilitate the development of independent work skills, and the specific type used in this study—educational applications—helps bridge the digital divide.

Keywords Education · Higher education · Digital educational resource · E-learning · Self-study · Language training · Modernization · Digitalization

JEL codes I20 · I21 · I23 · O15

S. B. Im (✉) · G. A. Jakupov
Uzbekistan State World Languages University, Tashkent, Uzbekistan
e-mail: imsb1985@mail.ru

S. M. Sultanova et al. (eds.), *Sustainable Development of Transport*, Lecture Notes in Networks and Systems 1366, https://doi.org/10.1007/978-3-031-88846-5_22

1 Introduction

Increasing requirements for the level of professional training of students of higher educational institutions set a task for teachers to modernize the educational process. In turn, the high pace of digitalization in all areas of life in today's society determines the need to increase digital literacy and instill skills in the use of information and communication technologies (ICT) for students of all fields, particularly future transport workers. The combination of these factors has created a need to introduce elements of digital education into the practice of training transport specialists in higher education.

One of the most productive ways to use ICT in the educational process is to use a digital educational resource (DER). A digital educational resource includes photographs, video clips, static and dynamic models, virtual reality and interactive modeling objects, cartographic materials, sound recordings, symbolic objects and business graphics, texts, and other educational materials necessary for organizing the educational process, presented in digital form [9].

This research aims to describe the methodological possibilities presented by DER in teaching Russian to students of transport specialties. These developments can be applied when teaching any other language or non-linguistic discipline because the developed principles apply to many academic disciplines.

In this research, the authors expect to solve the following problems:

- To explore the experience of foreign and Russian teachers in developing digital educational resources;
- To consider the existing classifications of DER and determine the most effective types of DER in teaching Russian language;
- To develop the structure of DER considering the features of learning Russian language by future transport workers;
- To present the use of interactive components of the DER for testing and self-testing students' knowledge;
- To create a system of exercises on Russian spelling, punctuation, vocabulary, and grammar to demonstrate the capabilities of the DER for learning different levels of language.

2 Materials and Methods

Several Kazakh researchers, including Yerahmetkyzy et al. [13], note positive aspects of the use of digital educational resources in the educational process. A study by specialists from the Polytechnic Institute of Braganca (Portugal) [8] obtained similar results for higher education. The authors consider DER an effective tool that accompanies traditional learning.

The research "Using Digital Educational Resources to Develop Self-reliance in Adolescents" by Akhmetov is devoted to the issues of developing independence and cognitive activity of students using digital educational resources [1].

The effectiveness of using DER in teaching a foreign language in a higher educational institution is noted by Shibaeva [12].

The COVID-19 pandemic created an urgent need to improve distance education methods. Due to the impossibility of face-to-face interaction between teachers and students, the emphasis has shifted towards independent education and the use of a computer or other remote communication means. In such conditions, the role of DER increases significantly. From serving as an auxiliary tool, a DER has evolved into the main method of interaction between students and the subjects they are studying. Sanz-Labrador, Cuerdo-Mir, and Doncel-Pedrera [11] studied the impact of the COVID-19 pandemic on the spread of digital educational resources as a means of education. Researchers note that the COVID-19 pandemic has become a serious threat to the quality of education due to the impossibility of learning in the traditional form. Simultaneously, the popularity of DER has increased significantly.

Issues of accessibility of DER and obstacles to their use are studied by Beyene, Mekonnen, and Giannoumis [3], as well as Rodés Paragarino and Gewerc [10]. The authors considered the social aspect of the restrictions associated with the use of digital resource centers, using Ethiopia and Latin American countries as examples. Similar issues of accessibility of digital education in the Russian-language educational space are covered by Alexandrov, Ivaniushina, and Simanovskiy. They highlight the concept of the digital divide and consider ways to overcome it [2]. The COVID-19 pandemic exposed a digital divide in teaching practices within the Republic of Uzbekistan.

Jokhan A. Chand, and others developed a DER using the latest advances in the field of AI [6].

Katkhanova considers DER as a subject of pedagogical design [7]. Valuable practical recommendations for designing digital access centers are presented by Gorlushkina and Gordeeva [4, 5].

During this research, the authors applied various methods, including description, comparison, statistical method, computer modeling method, and various diagnostic methods (e.g., exercises, surveys, and tests with single or multiple choice).

3 Results

DER can be classified based on various criteria. For example, according to the type of information presented, digital information centers with text, visual, audio, audiovisual, and combined information are distinguished. Based on the application's nature, there are DERs of teaching, training, and control types [12, p. 268]. However, most DERs combine the features of all types, having multimedia, which involves combining different ways of presenting information, and multifunctionality, combining means of providing information and testing and self-testing students' knowledge. The research proposes to classify DER according to the nature of information placement.

The first type is electronic catalogs. This type of DER is a collection of educational materials of various types within one website or application. In this case, links to educational materials or only information about them are provided, allowing students to independently find the necessary data. The director of the catalog is not the original author of the linked materials. For students to use this type of DER, they require access to the electronic catalog and the included sources.

Another type is educational portals. The educational portal combines information and interactive educational materials located on one website. Unlike electronic catalogs, educational portals provide full access to all educational materials that make up their content. The creators of the educational portal cannot post materials that violate copyright.

The advantage of electronic catalogs and educational portals is the ability to place many educational materials and update them as needed. Educational apps lack these benefits. As a rule, unlike the first two types, an educational application is devoted to only one subject or a block of subjects. To supplement the educational application with new materials, it is necessary to release a new version of it, which is essentially a separate program. However, the main advantage of educational applications is completely autonomous work in offline mode. Based on the features described above and the presence of a digital divide associated with access to the Internet, this research describes the features of using an educational application type.

Interactive tasks are the most important component of the DER from the point of view of monitoring student progress and self-testing of knowledge. They make it possible to reduce the load on the teacher through automatic checking tools and simulate the work of a student in the classroom in an online learning environment.

Developing a fully self-contained educational application that functions offline, supports multiple platforms such as Android and Windows, and offers extensive interactive user capabilities is possible with the RenPy development environment. By using RenPy, one can illustrate the interactive strengths of DERs in teaching the Russian language.

Exercise options are presented below.

Exercise 1. Single choice. Spelling.

Label test 1:

Menu:

"Choose the option with the correct spelling:"

"a) Альтернативный маршрут, пешеходный переход, перекресток":

call correct

"b) Альтэрнативный маршрут, пешеходный периход, перикресток":

"c) Альтернативный маршрут, пишеходный периход, перикресток":

"d) Альтэрнативный маршрут, пишеходный переход, перекресток."

Exercise 2. Test. Single choice. Punctuation.

Label test 2:

Menu:

"Choose the option in which the commas are placed correctly: Наземная дорожная разметка (1) нанесённая таким образом (2) чтобы создавать звук вовремя движения транспортных средств (3) по ней:"

"a) 1, 2, 3":

"b) 1, 2":

call correct

"c) 1, 3 ":

"d) 2, 3."

Exercise 3. True or false. Vocabulary. Synonyms.

label true_or_false1:

Menu:

"Пробка и выделенная полоса являются синонимами"

"True":

"False":

call correct

label true_or_false2:

Menu:

"Пробка и дорожный затор являются синонимами"

"True":

call correct

"False":

For automatic checking, each time a student selects the correct answer, the program calls a function that adds one point.

label correct:

$ mark_test = mark_test + 1

return.

The following tasks use hypertext markup.

Exercise 4. Multiple choice. Vocabulary.

Specify loanwords among the following terms:

Беспилотный Автомобиль, Выделенная Полоса, Бульвар, Двухсторонняя Проезжая Часть, Дорожная Диета, Зебра, Шлагбаум, Дорожная Развязка.

Correct answers: *автомобиль, бульвар, диета, зебра, шлагбаум.*

Exercise 5. Multiple choice. Morphology.

Find the nouns in the following sentence:

Альтернативный Маршрут—Шоссе, Которое Отделяется От Основной Магистрали И Снова Соединяется Через Некоторое Расстояние.

Correct answers: *маршрут, шоссе, магистрали, расстояние.*

Find the nouns in the following sentence:

Автомобиль Скорой Помощи—Транспортное Средство С Медицинским Оборудованием, Которое Перевозит Пациентов В Лечебные Учреждения, Например, В Больницы.

Correct answers: *автомобиль, помощи, средство, оборудованием, пациентов, учреждения, больницы.*

Find the nouns in the following sentence:

Автомобильная Дорога—Объект Транспортной Инфраструктуры, Предназначенный Для Движения Транспортных Средств.

Correct answers: *дорога, объект, инфраструктуры, движения, средств.*

Find the nouns in the following sentence:

Автомобиль Скорой Помощи—Транспортное Средство С Медицинским Оборудованием, Которое Перевозит Пациентов В Лечебные Учреждения, Например, В Больницы.

Correct answers: *скорой, транспортное, медицинским, лечебные.*

Exercise 6. Spelling. Fill in the missing letters.

Альт(1)рнативный М(2)ршрут, П(3)шеходный Пер(4)ход, Пер(5)кресток, Ис(6)кус(7)твенная Н(8)ровность.

Correct answers: *1—е, 2—а, 3—е, 4—е, 5—е, 6—_, 7—с, 8—е.*

Exercise 7. Punctuation. Put in a comma where needed.

Шумовая полоса—наземная дорожная разметка (1) нанесённая таким образом (2) чтобы создавать звук вовремя движения (3) транспортных средств по ней (4) сигнализируя о наезде (5) водителя на линию (6) например (7) когда водитель отвлёкся (8) или заснул за рулём.

Correct answers: *1, 2, 4, 6, 7.*

Exercise 8. Terminology. Match the following words so that they form terminological phrases.

Пешеходный –Автодорог А

Проезжая –Перехо Д

Магистральная –Полос А

Дорожный –Указате Ль

Кольцевая –Улиц А

Выделенная –Част Ь

Correct answers: *пешеходный переход, проезжая часть, магистральная улица, дорожный указатель, кольцевая автодорога, выделенная полоса.*

Thus, the structure of the interactive component of the digital educational resource can include various tasks, such as tests with single and multiple choice, yes/no questions, and exercises with hypertext. The capabilities of DER are not limited to the listed types of diagnostic tools. Their use depends only on a teacher's knowledge, creativity, and skills in pedagogical design. However, even the types of tasks above, which do not require the teacher to have extremely high-level knowledge of computer technologies, make it possible to cover almost all sections of linguistics required for teaching language to transport specialists: spelling, punctuation, lexicology, morphology, and knowledge of special terms. In general, digital educational resources have proven to be an effective tool for traditional and remote learning.

4 Conclusion

During their research, the authors formulated methodological foundations for building independent educational competence in teaching Russian to future transport specialists through self-education with a digital educational center. The effectiveness of DER technologies has been confirmed.

Despite the wide range of methodological tools offered by DER, they are more focused on helping the teacher than on helping study the subject independently. The teacher's role in the educational process remains significant despite the high level of digitalization. The search for the optimal balance between traditional and innovative forms of training, as well as the most organic ways of introducing digital learning into the educational process, remains promising for further research in this area.

References

1. Akhmetov SI (2022) Using digital educational resources to develop self-reliance in adolescents. Issues Jsm Educ Linguist 41(1):60–66. https://doi.org/10.52575/2712-7451-2022-41-1-60-66
2. Alexandrov DA, Ivaniushina VA, Simanovskiy DL (2017) Online educational resources for schoolchildren and the digital. Educ Stud Mosc 3:183–201. https://doi.org/10.17323/1814-9545-2017-3-183-201
3. Beyene WM, Mekonnen AT, Giannoumis GA (2023) Inclusion, access, and accessibility of educational resources in higher education institutions: exploring the Ethiopian context. Int J Incl Educ 27(1):18–34. https://doi.org/10.1080/13603116.2020.1817580
4. Gordeeva NO (2007) Features of teaching the design of digital educational resources. Sci Tech J Inf Technol Mech Opt 41:60–62
5. Gorlushkina NN, Gordeeva NO (2007) Methodology for teaching the design of digital educational resources. Sci Tech J Inf Technol Mech Opt 45:16–19
6. Jokhan A, Chand AA, Singh V, Mamun KA (2022) Increased digital resource consumption in higher educational institutions and the artificial intelligence role in informing decisions related to student performance. Sustainability 14(4):2377. https://doi.org/10.3390/su14042377
7. Katkhanova YuF (2010) Analysis of digital educational resources from an instructional design perspective. Prepodavatel XXI vek [XXI Century Teacher] 1(4):76–85
8. Morais C, Miranda L, Alves P (2015) The use of digital educational resources in the support to learning in higher education. EAI Endorsed Trans E-Learn 2(5):e5. https://doi.org/10.4108/el.2.5.e5
9. Polat ES, Bukharkina M. Yu, Moiseeva MV, Petrov AE (2009) New pedagogical and information technologies in the education system. Academa: Moscow, Russia
10. Rodés Paragarino V, Gewerc A (2023) Educational innovation, open educational resources, and gender in Latin American universities. Educ Sci 13(1):19. https://doi.org/10.3390/educsci13010019
11. Sanz-Labrador I, Cuerdo-Mir M, Doncel-Pedrera LM (2021) The use of digital educational resources in times of COVID-19. Soc Media Soc 7(3):1–19. https://doi.org/10.1177/20563051211049246
12. Shibaeva IV (2016) About using modern digital educational resources in teaching the French language in an institution of higher education. Scientific notes of V.I. Vernadsky Crimean Federal University. Philol Sci 2(68):267–273
13. Yerahmetkyzy Z, Shyryn A, Kulgildinova T, Lazzat A, Yerlan A, Ulbala I (2022) Teaching children's literature to school children through digital educational resources. World J Educ Technol: Curr Issues 14(5):1332–1342. https://doi.org/10.18844/wjet.v14i5.8093

A Solution to Language Problems in the Communication and Documentation of Local and International Employees in the Development of the Transport Sector

Oybek R. Achilov, Shokhista I. Mansurova, Zulfiya M. Sidiknazarova, and Rano M. Kiyasova

Abstract Uzbekistan is not only a supporter of the intensive transport sector development in Central Asia. Uzbekistan aims to connect Eastern and Western countries, from China to Europe, by reviving its historical role on the ancient Great Silk Road through advanced transit systems. The development of Uzbekistan's transport sector is a key public policy priority. Various organizations, including the Ministry of Transport and "Uzbekistan Railways" JSC, are actively involved in improving the country's transport infrastructure. Language plays a crucial role in creating a unified transport policy to ensure the delivery of high-quality services. This research aims to address linguistic challenges in international transport, such as misunderstandings, documentation errors, and communication barriers. To solve these issues, the authors applied cognitive-semantic analysis to create a three-step conceptual framework for understanding transport language. Systematizing this process with digital technologies could provide solutions for transport sector development. To systemically solve such problems, the authors studied cognitive-semantic properties of the language layer using frame analysis, comparative analysis, and methods of conceptual integration. As a result, the authors managed to form a three-step layer of concepts. The process begins by presenting words and terms in their basic form during the first stage. In the second stage, their meanings are generalized. In the third stage, a specific part relevant to the desired area is distinguished. Although this three-step language analysis

O. R. Achilov (✉) · S. I. Mansurova · Z. M. Sidiknazarova · R. M. Kiyasova
Tashkent State Transport University, Tashkent, Uzbekistan
e-mail: achilovoybek88@gmail.com

S. I. Mansurova
e-mail: umarovasabina20071980@gmail.com

Z. M. Sidiknazarova
e-mail: zulfiyasidi9@gmail.com

R. M. Kiyasova
e-mail: gapparovarano@gmail.com

S. M. Sultanova et al. (eds.), *Sustainable Development of Transport*, Lecture Notes in Networks and Systems 1366, https://doi.org/10.1007/978-3-031-88846-5_23

may initially seem challenging, if systematized using digital technologies, it could be proposed as a potential solution for advancing the development of the transport sector.

Keywords Transport sector · Reforms · Economic development · Intellectual systems · Concept · Frame analysis · Cognitive semantics · Conceptual integration · Foregrounding · Slots · Linguistic aspects · International transport services · International communication · Documentation · Advanced technologies

JEL codes A19 · A39 · B17 · C78 · C79 · F5 · R4 · Y1

1 Introduction

In today's world, technological advancements have surged, yet many challenges remain unsolved, particularly in social sectors where the role and significance of transportation are growing. These challenges are surfacing and becoming obstacles to development. In Uzbekistan, not only does the head of state prioritize the development of the transport industry, but several key organizations are also actively involved. These include the Ministry of Transport, the Ministry of Investment and Foreign Trade, Uzbekistan Railways JSC, Uzbekistan Airways JSC, Uzbekistan Airports JSC, the State Asset Management Agency, the State-Private Partnership Development Agency, the Antimonopoly Committee, and other relevant departments. Together, they are working to increase economic growth, ensure technical and operational safety, and elevate Uzbekistan's global position by enhancing its role in the transport sector. Nowadays, cooperation aims to achieve economic or technical advancements, foster mutual understanding in the political arena, and deliver services based on mutual agreements. In this political space, exchanging ideas through interstate communication and language tools is crucial in shaping prospects for mutual development. This research analyzes the role, specific features, and challenges of language tools in developing the transport sector and offers proposals for potential solutions.

2 Materials and Method

This research analyzes the cognitive-semantic aspects of linguistic units in the transport sector, considering that the most important features of the language are words and terms. The cognitive-semantic property refers to understanding the broader meaning behind a word's form. In the field of transport, many terms allow for multiple interpretations, and when these are organized into a systematic model, the desired properties can be represented in structured schemes. Several analytical methods are applied in this process, including Fillmore's frame analysis method, which connects form and content through cognitive mechanisms [1]. Fauconnier's conceptual integration method explores mental spaces [2]. Langacker's method focuses on cognitive domains [3]. Additionally, frame semantics analysis developed by Evans and Green [4] and reframe analysis developed by Bandler [5] are widely used in understanding these complex meanings.

Frame schemes developed by Ashurova and Galieva are very handy for text and term analysis [6]. Of certain interest are Djusupov's reframe analysis [7] and Rustamov's analysis of word-specific features [8]. In the field of transport, the analysis of words draws on the scientific research of several scholars, including Abdurakhmanov et al. [9], Aripov et al. [10, 11], Baxodirovich et al. [12], Bekzhanova et al. [13], Freimane et al. [14], Jumayev et al. [15], Kamenga et al. [16], Masharipov et al. [17], Prokopov et al. [18], Rasulov et al. [19], and Suyunbayev et al. [20].

Language problems in the transport sector can manifest in various ways, creating challenges for communication, safety, and efficiency. Some common language-related issues in the transport sector include the following aspects.

The main common type is a multilingual workforce. It can be analyzed using different languages in one working area. Thus, transportation staff may speak different languages in regions with diverse linguistic backgrounds. This can lead to challenges in effective communication and coordination. As a result, miscommunication among team members can result in errors, delays, or safety hazards.

Polysemy, the phenomenon where a single word has multiple meanings, is a significant aspect of language study. In various domains, polysemous words may be expressed differently, or conversely, the same concept may be referred to by different terms, leading to potential misunderstandings among language users or customers. In the transport sector, different regions or organizations may adopt varied terminology for identical concepts, which can create confusion among stakeholders. A viable solution to this issue is the standardization of terminology, which would improve communication and enhance mutual understanding within the industry [8].

The effectiveness of international business communication is manifested in the process that requires speed, especially in transport. This ensures correct and complete data exchange. In the case of international transportation, diverse languages, and regulatory frameworks may complicate cross-border operations.

The preparation of documentation and the declaration of vehicles used for importing and exporting international goods demonstrate the close connection between language and the transport sector. Conducting cargo transportation while adhering to the laws of various countries requires clear communication, further emphasizing this link. However, technical manuals, safety guidelines, and other essential documentation may only be available in certain languages, creating comprehension challenges for workers who are not fluent in those languages. To address this issue, translation services and the provision of multilingual materials are vital for ensuring smooth operations and avoiding misunderstandings. In addition to understanding laws, documents, and regulations, it is crucial in international activities to also grasp the customs and culture of the nations involved. Respecting and understanding cultural differences can strengthen business relationships and enhance cooperation. However, insufficient awareness of cultural nuances in communication can lead to misunderstandings or conflicts among transportation staff or between staff and passengers [3].

It is possible to carry out countless research and investigations on the specific features of each problem that has been mentioned above. The focus of this research is the analysis of words that are most relevant and serve as key sources for understanding the characteristics of language. To achieve this, it is essential to first provide a comprehensive overview of cognitive semantics, as it aligns with the research objective.

Cognitive semantics primarily concerns the conceptual nature of linguistic meanings and their relationship with conceptual structures that represent human knowledge and experiences.

Fillmore's theory of frame semantics, developed in 1982, is considered a highly influential theory in the field of cognitive linguistics. According to Fillmore, a frame is a structured form of knowledge that represents human experiences in specific contexts [1]. In essence, frames encompass intricate knowledge structures that include interconnected words and ideas. For example, "bus" is not just a means of transportation; it is associated with a range of concepts such as driver, passengers, parents, teachers, students, tourists, delegation representatives, guests, conversations, trips, maps, tickets, schedules, boarding, departure, cancellation, motor, transmission, brake system, or gasoline. As observed in the example of the concept of a bus, it is possible to identify many layers of meaning based on the knowledge structure within a single word form. The determination of meanings in the speech process depends on the specific context in which they are being used and is closely tied to the linguistic units that interact during speech.

The structural analysis of the concept of "bus" is carried out, indicating that a bus is a public transportation vehicle primarily used to transport school students and government employees to and from work and school, returning to general destinations in the evening. The analysis of speech acts (lexical units expressing it) and the associative connections of the word "bus" demonstrate its expression with the following linguistic units.

The associative links of the lexeme "bus" include motorbus, trackless trolley, motor coach, jitney, double-decker, charabanc, fleet, passenger vehicle, trolley coach, autobus, omnibus, public transport, window, roof, coach, trolleybus, school bus, and others.

The analysis of speech units and their associative directions in the concept of "bus" in speech makes it possible to open the following slots of the "bus" frame in Fig. 1 (i.e., it provides the opportunity to analyze its content):

1. Slot condition: New, old, broken, fixed, in service, and others;
2. Slot size: Minibus, double-decker, jitney, omnibus, and others;
3. Slot passenger: Kids, pupils, students, delegation members, tourists, prisoners, workers, guests, and others;
4. Slot wheels: four wheels, six wheels, eight wheels, twelve wheels, and others;
5. Slot aims: traveling, trips, visits, transportation, and others;
6. Slot fuel: diesel, gas, petrol, oil, elector power, and others;

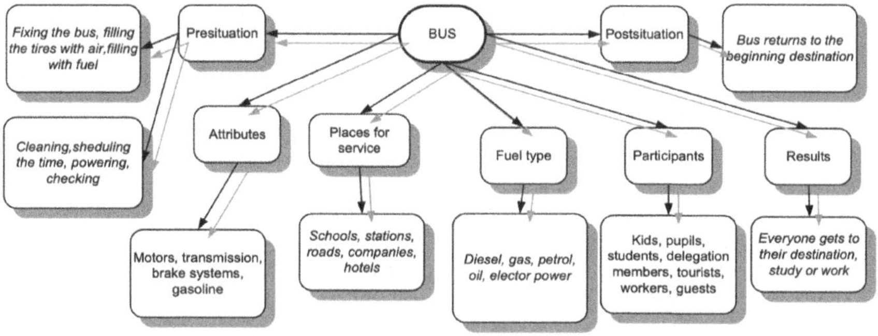

Fig. 1 Frame analysis of the word "bus." (*Source* Compiled by the authors based on [1, 4])

7. Slot order: last bus and first bus;
8. Slot color: red, green, blue, black, white, and others;
9. Slot drivers: with male or female drivers, driverless;
10. Slot agents: driver, passengers, masters, and others;
11. Slot attributes: motor, transmission, brake system, gasoline, and others;
12. Slot place: schools, stations, roads, companies, factories, hotels, and others;
13. Slot preparation: cleaning, scheduling the time, powering, checking, and others;
14. Slot pre-situation: fixing the bus, filling the tires with air, filling with fuel, and others;
15. Slot result: everyone gets to their destination, study, work, and others [6].

Thus, based on the frame analysis presented by Fillmore, the authors revealed that the word "bus" can encompass various knowledge structures [1]. An electrical conductor that makes a common connection between several circuits and, in this case, other words can be associated with it: computer, computing device, power plant, powerhouse, LAN, conductor, information processing system, computing machine, electronic computer, busbar, local area network, data processor.

The Fillmore's method is used to understand the meanings in the slot fillers of the concept of the bus presented in the above example. The result of the frame analysis is presented in Fig. 2.

Turning to the next example, the authors carried out a structural analysis of the concept of "ballast," indicating a heavy substance (such as rocks or water) placed to improve stability and control (as of the draft of a ship or the buoyancy of a balloon or submarine). The analysis of speech acts (lexical units expressing it) and the associative connections of the word "ballast" demonstrate its expression with the following linguistic units.

The associative links of the lexeme "ballast" are sailing ballast, ballast car, electronic ballast, combined ballast, over-ballast, sub-ballast, under-ballast, cultural ballast, homegrown ballast, water ballast, structural ballast, moorum ballast, cinder ballast, brickbat ballast, and others.

The analysis of speech units and their associative directions in the concept of "ballast" in speech opens the following slots of the "ballast" frame in Fig. 2:

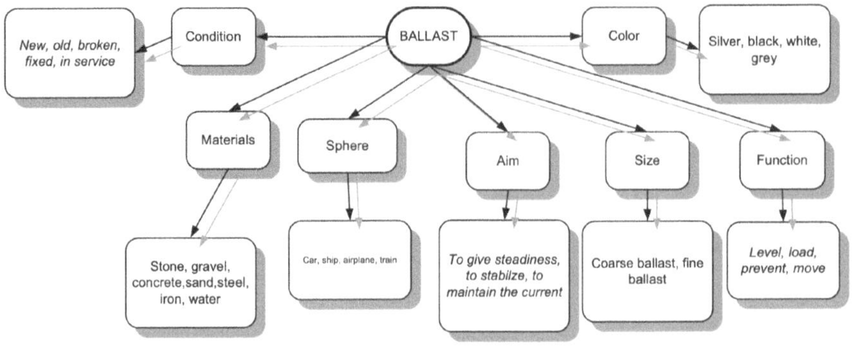

Fig. 2 Frame analysis of "Ballast." (*Source* Compiled by the authors based on [1, 4])

1. Slot condition: new, old, broken, fixed, in service, and others;
2. Slot materials: crushed stone, washed gravel, bank run (unwashed) gravel, torpedo gravel, heavy concrete, sand, steel or iron, water;
3. Slot sphere: car, ship, airplane, train, and others;
4. Slot aims: to give steadiness, to provide stability, to maintain the current, and others;
5. Slot size and section: coarse ballast (typically larger particles, such as 20–40 mm or 40–60 mm), fine ballast (smaller particles, such as 5–20 mm), and others;
6. Slot color: silver, blue, black, white, and others;
7. Slot functions: provides a leveled bed, transfers the load from sleepers to subgrade, holds the sleepers in a firm position, prevents the longitudinal and lateral movement, offers good drainage, and others;
8. Slot attributes: aluminum, silicon, and others;
9. Slot place: ocean, sea, station, ports, roads, and others [6].

Thus, the ballast can have the following meanings:

1. Water taken into a ship to maintain stability and balance in water transport;
2. Rail track ballast serves as a bed for rail tracks and provides drainage and strength for heavy loads carried by trains;
3. Ballast is a weight, usually of rubber or sand, used in an aircraft or ship to attain the desired balance when the center of gravity is not within limits or is not at the location selected by the operator;
4. A device used to provide the starting voltage or to stabilize the current in a circuit (as of a fluorescent lamp) [21].

In the frame of "ballast," we can clearly understand that the first form of the word does not convey the general or needed semantic meaning. The top level of the linguistic unit is the sign to go deeper or search the lower level. Therefore, it is highly important to clarify the term frame in linguistics. A frame is a hierarchical structure of linguistic information. It consists of upper and lower levels. The top level is the name of the frame. The lower level consists of terminals (English slots and sub-slots) and conveys specific information about the situation under consideration. All terminals that make up the frame are not independent; they are closely related and interconnected with others. Each terminal indicates the terms and conditions of a particular situation, which is characterized by a range of properties and qualities. Some of these features are apparent, while others are hidden. During frame analysis, hidden components must be decoded and interpreted. From this point of view, it is appropriate to say that frames create new sensations. Additionally, the frame is not a stable structure. It responds to changes with changes in the conceptual worldscape.

3 Results

To understand the meaning and achieve analytical results for lexical units, Fauconnier's conceptual integration method has been applied [2]. This method involves three key spaces:

- Input space is the entry space where individual elements or concepts are introduced for consideration;
- Blending space is the space where different thoughts, ideas, and concepts are mixed and integrated, allowing for new interpretations and connections to emerge;
- Output space is the exit space, representing the result or outcome of the blending process, where the final, integrated meaning is derived.

Thus, on the input stage, the word "bus" may have two different meanings: (1) transport means or (2) conductor. As the process moves into the next stage of conceptual integration, various concepts that arise in the human mind can be considered. These might include the following:

1. Driver, passengers, parents, students, tourists, delegations, guests, conversations, trips, maps, tickets, schedules, boarding, departure, cancellation, motor, transmission, brake system, gasoline, and others;
2. Conductor, computer, computing device, power plant, powerhouse, LAN, information processing system, computing machine, electronic computer, busbar, local area network, data processor, and others.

At the output stage, the addresser, based on the processed and blended text information, will have a clearer understanding of the given term or concept. This clarity will allow the addresser to make an informed decision, selecting the correct meaning depending on the context or goal of the communication.

In Fauconnier's conceptual integration method, the word "bus" in the entry space exists in two forms, complicating its informational acceptance process. This complexity arises from the human cognitive gap, where various layers of individual perspectives, prior experiences, and knowledge structures influence interpretation. These layers are integrated within the conceptual blending process, making it harder to resolve the meaning without further analysis. As a result, a comprehensive analysis is required. Words, phrases, lexical units, and relevant contextual information are integrated into the blending space. Through this process, the understanding of the word "bus" is refined by resolving ambiguities. As the concept moves from the entry space to the exit space, the meanings are distinguished, analyzed, and extracted based on the cognitive and contextual framework established in the model.

In the definition of lexical units, there are other difficult terms that pose a similar challenge, including the word "fare." In the entry space, it is the money that one pays for a journey (e.g., in a bus, train, or taxi). In the conceptual integration space, there are meanings such as get along, get on, get by, do, cope, manage, turn out, pan out, or proceed. In the exit space, it reflects the exact purpose for which it is used [8, 22].

In the conceptual integration space, it is possible to express meanings such as a brace, bracket, counterbalance, counterweight, sandbag, stability, and stabilizer. However, the output specifically identifies the purpose for which the word is used.

Understanding this transport sector jargon is essential for effective communication within the industry. It ensures that professionals can convey information accurately and efficiently.

Thus, the semantic aspect plays a crucial role in comprehending the text. The frame analysis can be outlined as a sequential procedure involving the following:

1. Recognizing verbal cues that convey essential conceptual frames;
2. Unraveling these frames through semantic, associative, metaphoric, and contextual links;
3. Activating knowledge structures, contextual details, and propositional functions;
4. Interpreting textual information, encompassing generalization, drawing inferences, and extracting knowledge based on verbal cues.

Every phase of the analysis contributes to elucidating cognitive content. It is feasible to unveil concealed frame elements and their associations throughout the frame analysis process, allowing for the derivation of results and conclusions.

Based on these points, the following conclusions can be drawn. Providing language training for staff involves implementing programs or initiatives aimed at improving the language proficiency of employees within an organization. This training can encompass various aspects of language learning, including vocabulary building, grammar instruction, speaking practice, listening comprehension, and writing skills development. The goals of language training for staff may vary depending on the specific needs and objectives of the organization. For example, it could enhance communication skills for better collaboration among multicultural teams, prepare employees for international assignments or interactions with clients from different linguistic backgrounds, or meet regulatory requirements in certain industries.

Implementing standardized terminology involves establishing a uniform set of terms, phrases, or vocabulary within a particular domain or industry to ensure clear and consistent communication among stakeholders. This process is crucial for promoting accuracy, efficiency, and comprehension in various contexts (e.g., healthcare, technology, finance, and manufacturing).

Ensuring the availability of translated documentation involves accurately translating all relevant materials (e.g., manuals, guides, policies, and instructions) into the necessary languages. This ensures that the diverse needs of customers, employees, and stakeholders are effectively met.

Utilizing technology for real-time translation and communication involves leveraging advanced tools and platforms to facilitate seamless and instantaneous language translation during live interactions. This enables individuals who speak different languages to communicate effectively in real-time.

Addressing language problems in the transport sector involves implementing strategies and solutions to overcome communication barriers that arise due to linguistic diversity among passengers, staff, and stakeholders. By addressing these language challenges, it is possible to significantly enhance efficiency, safety, and overall customer satisfaction.

4 Conclusion

In recent years, large-scale measures have been implemented in Uzbekistan to develop the field of transport and transport communications, ensuring a high level of transport safety, improving the management system in the field of transport, and providing training qualified specialists for the field.

To fundamentally improve the state management system in the transport sector, enhance the republic's investment attractiveness and export potential, ensure the strategic development and stable operation of transport communications, and address the five priority development areas of the Republic of Uzbekistan from 2017 to 2021, a comprehensive strategy was established. This strategy is the foundation for the subsequent development plan, "New Uzbekistan for 2022–2026," which continues to build on these objectives [23, 24]. To summarize and address language-related challenges, the transport sector can implement several strategies, including the following:

- Providing language training for staff;
- Implementing standardized terminology;
- Offering multilingual customer support;
- Ensuring the availability of translated documentation;
- Utilizing technology for real-time translation and communication;
- Addressing language problems in the transport sector to enhance efficiency, safety, and overall customer satisfaction.

Considering the solutions mentioned above, language can be recognized as a crucial guideline in the development and implementation of a unified transport policy. It is key in driving modern economic trends and supporting the country's ongoing reforms.

References

1. Fillmore C (1982) Frame Semantics. In linguistics in the morning calm. Hanshin Pub. Co., Seoul, South Korea, pp 90–111
2. Fauconnier G, Turner MB (1998) Conceptual integration networks. Cogn Sci 22(2):133–187
3. Langacker RW (1987) Foundations of Cognitive Grammar. Theoretical prerequisites, vol 1, Stanford University Press, Stanford, CA
4. Evans V, Green M (2006) Cognitive linguistics: An introduction. Routledge, New York, NY
5. Bandler R (1983) Reframing: neuro-linguistic programming and the transformation of meaning. Meta Publications
6. Ashurova DU, Galieva MR (2016) Text linguistics. Tashkent, Uzbekistan. Turon-Iqbol
7. Dzhusupov NM (2013) Cognitive-stylistic approach to studying the title of a literary text (based on the english language fiction). Issues Cogn Linguist 2(035):47–61
8. Rustamov I (2020) Ethnic stereotypes in anecdotes. Int J Sci Technol Res 9(03):622–623. http://www.ijstr.org/final-print/mar2020/Ethnic-Stereotypes-In-Anecdotes.pdf. Accessed 2 Aug 2024
9. Abdurakhmanov A, Khudayberganov S, Khusenov U, Jumayev S (2023) Mathematical model of criteria for a comprehensive assessment of the state of railway transport infrastructure. E3S Web Conf 402:06006. https://doi.org/10.1051/e3sconf/202340206006
10. Aripov N, Sadikov A, Ubaydullayev S (2021) Intelligent signal detectors with random moment of appearance in rail lines monitoring systems. E3S Web Conf 264:05039. https://doi.org/10.1051/e3sconf/202126405039
11. Aripov N, Suyunbaev S, Azizov F, Bashirova A (2021) Method for substantiating the spheres of application of shunting locomotives at sorting stations. E3S Web Conf 264:05048. https://doi.org/10.1051/e3sconf/202126405048
12. Baxodirovich DB, Kobiljonovich SK, Mansuralievich SS (2020) Modeling of unproductive losses in the operation of a sorting station. Solid State Technol 63:6–19
13. Bekzhanova SE, Issina BM, Zhaksybaeva AZ, Suyunbayev SM (2020) Introduction of a single intellectual management system and automation of production processes on railway transport. J Adv Res Dyn Control Syst 12(6SI):706–712
14. Freimane J, Mezitis M, Mihailovs F (2016) Maneuver movements' safety increase using maneuver locomotive identification and distance control. Procedia Comput Sci 104:375–379. https://doi.org/10.1016/j.procs.2017.01.148
15. Jumayev S, Khudayberganov S, Achilov O, Allamuratova M (2021) Assessment criteria for optimization of parameters affecting to local wagon-flows at railway sites. E3S Web Conf 264:05022. https://doi.org/10.1051/e3sconf/202126405022
16. Kamenga F, Pellegrini P, Rodriguez J, Merabet B (2021) Solution algorithms for the generalized train unit shunting problem. EURO J Transp Logist 10:100042. https://doi.org/10.1016/j.ejtl.2021.100042
17. Masharipov M, Rasulov M, Suyunbayev S, Jumayev S, Bekmurodov S (2023) Establishing the impact of empty freight trains on the capacity railway lines. E3S Web Conf 431:08021. https://doi.org/10.1051/e3sconf/202343108021
18. Prokopov V, Bohomazova H, Prokhorov V, Kalashnikova T, Golovko T (2021) Constructing a model for the automated operative planning of local operations at railroad technical stations. East-Eur J Enterp Technol 3(3–111):32–41. https://doi.org/10.15587/1729-4061.2021.233673
19. Rasulov MX, Suyunbaev ShM, Masharipov M (2020) Research of development prospects of transportation hub in JSC Umc. J Tashkent Inst Railw Eng 16(3):71–77
20. Suyunbayev S, Khusenov U, Khudayberganov S, Jumayev S, Kayumov S (2023) Improving use of shunting locomotives based on changes in infrastructure of railway station. E3S Web Conf 365:05011. https://doi.org/10.1051/e3sconf/202336505011
21. Ballast (2024) Merriam webster dictionary. https://www.merriam-webster.com/dictionary/ballast. Accessed 2 Aug 2024

22. Suyunbayev SM, Khudayberganov SK, Svetashev AA, Jumayev SB, Mikhayeva DB, Olimov AR (2021) Method of determining the minimum required number of sorting tracks, depending on the length of the group of wagons. Revista Geintec-Gestao Inovacao e Tecnologias 11(2):1941–1960
23. President of the Republic of Uzbekistan (2019) Decree on measures to radically improve the system of state management in the sphere of transport (1 Feb 2019 No. DP-5647). Tashkent, Uzbekistan. https://lex.uz/docs/4194115. Accessed 2 Aug 2024
24. President of the Republic of Uzbekistan (2022). Decree on the development strategy of the New Uzbekistan for 2022–2026) (20 Jan 2022 No. DP-60). Tashkent, Uzbekistan. https://lex.uz/uz/docs/6968143. Accessed 2 Aug 2024

Investigating Linguistic Aspects of Terminology in the Automotive Sector

Lola A. Rikhsieva

Abstract The research aims to emphasize the importance of translators being aware of synonyms within the terminology of the automotive industry in English and knowing the differences in their usage across various countries. Nowadays, linguistic issues are becoming relevant for translators in the scope of terminology. Translators working with specialized terminology should be cautious of false cognates, which can lead to misunderstandings during translation. The research also determines the optimal method of the most adequate translation of terms translated from English and Russian by transliteration or borrowing. Additionally, the research reveals that English terms in the automobile industry share the same grammatical structures as any general lexical-semantic group. New terms are a significant source of language development. They contribute to the enrichment of glossaries. Foreign experience is crucial in the advancement of each field. Therefore, it is important to focus on the key linguistic features when training translators in this field.

Keywords Terminology · Terms · Automotive industry · Synonyms · Translation

JEL codes H75 · Y800

1 Introduction

Terminology is one of the important objects of research because it reflects the level of development of various branches of knowledge. As a branch of linguistics, terminology aims to study various connections of language with various sciences. It emerged as an interdisciplinary field at the intersection of several disciplines, primarily linguistics, logic, and semiotics, encompassing theoretical and applied terminology [1]. Like many terms that have long entered into scientific use, the lexeme term has many definitions. Therefore, it makes sense to determine the signs that are relevant to the scientific concept behind it. The analysis of scientific works,

L. A. Rikhsieva (✉)
Uzbekistan State University of World Languages, Tashkent, Uzbekistan
e-mail: lolarikhsieva@uzswlu.uz

S. M. Sultanova et al. (eds.), *Sustainable Development of Transport*, Lecture Notes in Networks and Systems 1366, https://doi.org/10.1007/978-3-031-88846-5_24

educational literature, and various lexicographical sources shows different definitions of the concept term, even in lexicology and terminology. On the one hand, such diversity indicates the object's complexity. This necessitates selecting the most general formulations and justifying the working definition of the term used to address the research problems [2]. The idea that the functioning of terms is limited to special communication firmly comes to the fore in the 20th century. In the semantization of a lexeme, terms are words that are accepted to designate specific concepts within a particular field or profession. Nowadays, there are three thousand more definitions of the concept of "term." However, the discussion of one universally recognized definition continues. There is still no unified agreement on this issue. On the one hand, this is explained by the insufficient development of the terminology system. On the other hand, the definitions available in the literature do not always fully reflect the essence of the concept of "term" [3]. The purposeful nature of the appearance of terminological units is called among the main features characterizing the specificity of the term. The main difference between the formation of terms and the formation of general language is that terms are not formed in the proper sense of the word and do not appear. They are created and invented as their necessity is realized. This famous statement by Vinokur has been commented on and developed by terminologists many times [4]. The term is the basis of scientific communication; it ensures accuracy, clarity, and understanding of scientific thought. By conveying conceptual information, the term plays an important role in communication processes. The purpose of writing special texts is to exchange scientific and technical information contained in the terms.

2 Literature Review

The research considers linguistic aspects of terminology in the automotive sector. Over the last decade, linguistics saw an increasing interest in the linguistic analysis of the special vocabulary of certain fields of knowledge. In the 20th century, the world transport system developed greatly. All transport systems in developed countries are characterized by the high level of development of Western European countries, the USA, the former Russian Empire (USSR), and Japan. Until the beginning of the 20th century, goods and passengers were transported in Uzbekistan mainly by railways, automobiles, camels, and horses. At the end of the 1990s, the transport industry was created with the construction of automobiles and large industrial enterprises. The material and technical base of the republic's transport somewhat improved. The terminology of the transport sector began to be formed. The development of the automotive industry in Uzbekistan has led to the emergence of numerous scientific publications and dictionaries on the studied topics. The contradictory nature of information in this kind of literature causes inconsistencies and inaccuracies in translating specialized texts and technical documentation. It also causes significant challenges in communication between specialists in the transport and automotive sectors. This research highlights the lexical-semantic field of automotive terminology in English and Uzbek languages, identifies characteristics of terms that do not differ from the general vocabulary, and carries out an interlingual comparison in translating the terms from English into Russian and Uzbek. The research also analyzes the history of the Uzbek automotive industry and the terminology formation and identifies translation difficulties. The material for investigating the specifics of the linguistic aspects of automotive terminology was the texts of scientific works, particularly special dictionaries. The theoretical and methodological basis of this research includes the works of Uzbek and foreign linguists in the sphere of terminology, including the works by Averbukh [1], Dadaboev [5], Danilenko [6], Leychik [7], Newmark [8], Novikov [9], Serbinovskaya [10], and Tatarinov [4]. Translation studies of terms are noted in the works of Proshina [11] and Bobyreva [12]. Word formation and lexical semantics of terms are applied in the works of Kadyrbekova [13]. The main sources of lexical material and analysis were the "Dictionary of automotive engineering" by Goodsell [14], used to understand the meanings of English terms, and the textbook for Uzbek specialists of the automotive sector "*Avtomobillar texnik ekspluatatsiyasi*" [Technical operation of automobiles] by Normirzaev and Mannonov [15], used to give an adequate translation of the terms from English into Uzbek. The author also used 15 scientific works related to terms in English, Russian, and Uzbek.

3 Materials and Method

Terms, terminological systems, and terminology have always been the subject of consideration by linguists. The specific features of terminological issues are studied in the scientific works of Akhmanova [16], Averbukh [1], Bobyreva [12], Dadaboev [5], Danilenko [6], Kadyrbekova [13], Leychik [7], Newmark [8], Novikov [9], Proshina [11], Serbinovskaya [10], Tatarinov [4], and Temmerman [2].

Nevertheless, there remain undisclosed problems related to the term's nature, which should be studied in accordance with the trends of today's linguistics. The use of terms in all sectors begins with the numerous studies on the problems of terms, terminology, and terminological systems. These problems were defined in different ways. The formulation of term definitions, term descriptions, the development of solutions to associated issues, and, ultimately, the establishment of the science of terminology were all prompted by the definition of the term as a linguistic object and its methodical study. In the author's view, the term is a necessary component of the general language system due to its linguistic character. It is distinct from other word categories in that it clearly and concisely represents concepts related to science, technology, or other professions.

According to Akhmanova, linguistic terminology is more complex than a basic vocabulary list. Linguistic terminology is a distinct semiological system rather than just a straightforward collection of terms like any other terminology [16]. Terminology only becomes apparent when a science achieves the pinnacle of its growth. The term becomes recognized only after a certain notion gains a precise scientific representation. These days, terminology plays a unique role in the development of worldwide scientific language. Scientific research is influencing the development of terminology across various sectors. A significant contribution has been made by Russian scientists, such as Serbinovskaya, whose study introduced the concept of a terminological system. It states that a terminological system comprises numerous connected and intersecting terminospheres that unify words, forming the system's core [10]. As the nuclear unit of the terminological system, there are terms that have features such as comprehensibility, precision, unequivocalness, definitiveness, and systematicity. Other special nominative units forming the periphery of the terminological system, which cannot fully meet the requirements for terms, are distinguished.

The lexical units that have the status of terms in the automotive industry are located at the core of the terminology system. They are considered pure, professional terms. In the periphery, there are lexical units of general usage that have been terminated. They usually have a metaphorical character of the meaning. Terms have the same characteristics as any lexical-semantic group of generally used lexical units such as synonymy, polysemy, antonymy, and others. There are enough shortcomings in the terminological system. There is no generally accepted definition of terms. Terms are used in the form of polysemy, homonomy, and synonymy.

Polysemantic terms cause considerable problems and misunderstandings in the translation process. On the one hand, this requirement seems reasonable. On the other hand, it contradicts the rules of common language. The different terminological systems are enriched by polysemantic terms, and their correct translation depends on the context. For example, in the field of transportation, the English term "vehicle" is used to mean a type of overland vehicle that is transported from one place to another, as well as "a bicycle," "a cart," and "a sled."

The use of synonyms in terminology often causes controversy among researchers. In common literary language, each word of a synonym string can have only one meaning in a given context. In this case, it is appropriate to use a synonym. According to Novikov, the peculiarity of the term is that regardless of the context, it has only one meaning. Therefore, synonyms of one concept are inappropriate in terminology. Synonyms are one of the most important categories in the general literary language. They are less important in terminological systems because they cannot show emotional-expressive features. As synonyms are disputable points in terminology, translators in certain fields often deliver messages of incorrect translations in the process. It is undeniable that many English terms in the automotive sector have synonyms. These terms should be translated with a single, precise meaning to ensure clarity and accuracy. The examples are as follows:

- Boot, trunk, hatch, tailgate, liftgate—*bagaj*;
- Vehicle frame, chassis—*shassi*;
- Motor oil, engine oil, engine lubricant—*motor moyi*;
- Bulbar, push bumper—*himoyalovchi qurilma*;
- Dashboard, fascia—*uskunalar paneli*;
- Muffler, silencer—*glushitel*;
- Parking brake, handbrake, emergency brake—*gidravlik qo'l tormozi*.

Novikov states that the terms included in the synonymy series are often redundant, which makes it challenging to master the terminology and hinders the presentation of information and its accurate reception [9]. Other scientists, such as Danilenko, emphasize that using synonyms in terminology can be beneficial. It allows for word choice flexibility and helps prevent repetition in the text [6]. Another feature of the linguistic aspect of terms, put forward by Leychik, emphasizes that terms, from a linguistic perspective, function as lexical units within terminological systems in terms of both structure and semantics [7]. In the first case, terms are defined as words, which can be categorized into individual words, derivatives, compounds, and abbreviated structures. However, another perspective suggests that units from other parts of speech should also be classified as terms. In the language of science, namely in terminologies of certain sectors, it is possible to freely use verbs, adjectives, adverbs, and abbreviations as terms if they express special concepts and meet the other requirements for terms [13]. Accordingly, the English terms in the automotive sector have the same morphological structural components as any lexical-semantic group of generally used lexical units, for example:

- Individual words without any affixation: piston, squat, stoke, strut, sump, yaw;
- Derivative terms with affixation such as prefix and suffix: oversteer, antifreeze, trafficator;
- Compound terms operating as individual lexical units: bulbar, crankshaft, floorpan, flywheel, valvetrain;
- Verb terms: dive, jounce, yaw;
- Poly lexemes containing two components, such as noun + noun or adjective + noun: gas pedal, motor oil, valve float, helical gear, hydraulic lifter;
- Abbreviation: ABS (anti-lock braking system), HP (horsepower), RKS (remote keyless system), TCU (transmission control unit).

4 Results

The regional use of the language (dialect) is an important factor in the use of synonyms in terminology. For instance, in the automotive industry, English terms differ between British and American dialects. The author proposes using the following model "term (BE) British-English + (AE) American-English = (UzT) Uzbek Translation" to clarify:

- Accumulator + battery = *akkumulyator*;
- Accelerator + gas pedal = *gaz pedali*;
- Bonnet + hood = *kapot*;
- Cubby box + glove compartment = *bardachok (barqutisi)*;
- Dual carriageway + divided highway = *ajratuvchi bo'lak (yo'lning yonma-yon joylashgan qatnov qismlarini ajratuvchi)*;
- Dynamo + generator = *generator*;
- Estate car + station wagon = *kuzov*;

- Flyover + overpass = *ko'prik (yo'l o'tkazgich)*;
- Indicator + turn signal = *ogohlantiruvchi-qayilish signali*;
- Gearbox/manual + transmission/stick shift = *kuch uzatma*;
- Number plate + license plate = *raqam belgisi*;
- Boot + trunk = *bagaj*;
- Lorry + truck = *katta yuk tashuvchi transport*;
- Petrol + gas = *benzin yoqilg'isi*;
- Saloon + sedan = *avtomobil saloni*;
- Tyre + tire = *shina*;
- Wing + fender = *qanot*;
- Windscreen + windshield = *peshoyna.*

There is a difference between driving a car in Britain and America. The British predominantly drive cars with manual transmissions, while Americans mainly use cars with automatic transmissions. This difference is reflected in the etymological origins of the terms, with "gearbox" and "manual" used in British English and "transmission" and "stick shift" in American English. Thus, translators should consider the differences between British and American English terms and their translations in the automotive industry. For example, in the US version, the external component is called the hood; in British English, it is the bonnet. This term is translated into Russian as *kanom.* The Russian word *kanom* is an acceptable and adequate translation. Thus, the direct translation from English to Uzbek may create some misunderstanding: hood—*kapyushon* or bonnet—*ayollar shlyapasi.* Field translators should beware of false cognates [17]. They include words that are similar in form but very different in meaning, as well as words that are not entirely opposite despite having common elements in their semantics [12]. For example, "gas tank" should be translated as *benzobak yoki yoqilg'i baki*, "glove compartment"—*bardachok(bar qutisi)*, "controller"—*o'tkazgich yoki datchik*, not as *gazoviy tank*, *qo'lpoqlar uchun quti*, and *kontrolyor*, because logical nonsense is visible in such translations.

The difficulties in translating car components and spare parts into Uzbek were evident even before Uzbekistan began producing cars and established its automobile industry, which does not have a long history. During the years of independence, enriching the Uzbek language's vocabulary was significantly influenced by the rapid formation and development of entirely new sectors for the national economy.

In 1996, the joint venture "UzDEU Avto" was launched in Asaka, Andijan Region, in cooperation with the South Korean Daewoo automobile corporation. By 1999, the Uzbek-Turkish joint venture "SamKochAvto" started producing buses and trucks in Samarkand. Currently, production of Isuzu buses and MAN trucks has begun in collaboration with Japanese and German car companies. Following the integration of the giant "UzDEU Avto" joint venture into the American car manufacturer General Motors, the goal is to produce new car models that meet contemporary requirements, offering high speed, comfort in all aspects, and fuel efficiency. As a result, the terminology of this field in the Uzbek language has been radically improved. Uzbek terminology (e.g., Tiko, Damas, Nexia, Lacetti, Matiz, Spark, Epica, Captiva,

Malibu, and Kobolt (Cobolt)) is even richer due to the terms meaning of the commercial name of cars put on the assembly line in the country [5]. As the automobile industry in Uzbekistan does not have a long history and there was no established automobile industry in the territory, cars and their spare parts and components were often imported from the former Russian Empire [17]. Thus far, the terms used in the field of transport have either been translated from Russian into Uzbek or Russian terms have been adopted as borrowings, for example, brake shoe—*tormoz kolodkasi (тормозная колодка)*, steering wheel—*rul' (руль)*, bonnet or hood—*kapot (капот)*, estate car or station wagon—*kuzov (кузов)*, tyre or tire—*shina (шина)*. Uzbekistan began to export cars since 1996. Attention has increasingly been paid to using terms in the pure Uzbek language. Many publications in this field contribute significantly to the accurate translation of automotive terminology. Using the educational and methodological aid of the subject "*Avtomobillar texnik ekspluatatsiyasi*" [Technical operation of automobiles] by Normirzaev and Mannonov for students of the transport institute, the author analyzed the translation process and provided an acceptable and adequate translation of the terms from English into Uzbek [15], including the following:

- Gearbox/manual or transmission/stick shift—*коробка передач* in Russian and *kuch uzatma* in Uzbek. However, the Russian version is still actively applied in speech;
- Crankshaft: *коленчатый вал* in Russian and *tirsakli val* in Uzbek.The Russian and Uzbek versions differ slightly in terms of word choice. In this case, the first morphological component adjective is expressed using a human limb in a metaphorical sense. The lower limb is in Russian, and the upper limb is in Uzbek.

To deliver adequate and comprehensible information when switching from one language to another, translators in any sector should be aware of the classifications of the translation of terms. Peter Newmark clarifies Paepcke's approach to distinguishing technical language levels, such as scientific, workshop, daily usage, and promotion or sales, noting that these distinctions are not appropriate for all terms in every sector. Instead, he suggested academic, professional, and popular levels based on medical terminology. In this framework, Latin words in the medical sector belong to the academic level, terms used by specialists are categorized as professional, and familiar alternative terms used by non-specialists fall under the popular level [8]. According to the author's experience in translation studies, the essential approach of the automotive sector involves using translation methods that pertain to the classification of terms based on levels of nomination corresponding to their meaning. The absolute correspondence level classifies terms that are clear from their nomination. The partial correspondence level includes terms that require explanation because their meaning is not immediately apparent from their names. The non-correspondence level involves terms whose meanings do not match their nominations at all, necessitating detailed explanations. According to the analysis, it follows as a logical consequence that in the automotive sector, the following three types of approaches are revealed by the scope of Don Goodsell's "Dictionary of automotive engineering." This dictionary clarifies the nomination of terms with absolute correspondence levels, which are clear from their names, such as:

- Clarifying nomination of absolute correspondence level terms, which is clear from its name (e.g., accelerator—the depression of the pedal causing the vehicle to accelerate, antifreeze—lowers the freezing point of a liquid, accumulator–rechargeable electrical storage battery, exhaust silencer—reduces and adjusts the noise of the exhaust system, regulator—device stabilizing or controlling the current and voltage in a vehicle electrical system);
- The partial correspondence level terms requiring explanation where necessary (e.g., caliper—in a disc brake system, the mechanism that brings brake pads to bear on the disc by clamping or pinching action, fascia—panel or molding immediately below the windshield on vehicle interior, usually facilitating mounting of instruments, air vents, and accessories);
- Non-correspondence level terms without any doubt requiring explanation all time in context because their semantics have metaphoric features (e.g., beaver tail—a ramped or lowering rear section of a low-bed trailer or low-loader; a *donut*—a temporary *spare tire* (*donut* tires are much smaller than standard tires); donkey engine—a small auxiliary engine, usually for driving ancillary equipment or services when the main engine is stopped; water jacket—a part of the cylinder block and head of an engine that encloses the cavity through which cooling water flows) [14].

This approach, which involves dividing terms into different categories, provides an opportunity to handle terms more effectively by defining the appropriate translation method and increasing accuracy in conveying messages from the source language to the target language. In most cases, the translation methods of transliteration and borrowing apply to absolute correspondence level terms, as these terms are clear from their names and do not require any additional description when translated from English into Uzbek. One of the ways is a transliteration, for example, antifreeze—antifriz, accumulator—akkumulyator, regulator—regulyator. Another way is borrowing from Russian: accelerator—pedal and exhaust silencer—glushitel. The partial and non-correspondence level terms tend to be translated in a descriptive manner. Descriptive translation involves using a description to convey the meaning of a term from the source language, rather than translating it directly according to its current definition: caliper—*tormoz supporti, tormoz tizimining asosiy elementlaridan biri bo'lib, tormozlash vaqtida tormoz kolodkasi diskga siqilishini va shu bilan avtomobilni to'xtatishini taminlaydi;* fascia—*avtomobilning ichki qismidagi old oynaning to'g'ridan-to'g'ri ostidagi uskunalar paneli bo'lib, odatda jihozlar, ventiliyatsion teshiklari va aksessuarlarni o'rnatish uchun mo'ljallangan qoplama;* beaver tail—*yuk ko'taruvchi treyler, yuk ko'taruvchi treylerning egilgan yoki tushirilgan orqa qismi*; a *donut—vaqtinchalik zaxira shinasi, standart shinalarga qaraganda ancha kichikdir*; donkey engine—*kichik yordamchi dvigatel, odatda asosiy dvigatel to'xtatilganda yordamchi uskunalarni boshqarish uchun mo'ljallangan*;

water jacket—*silindr bloki, sovutish suvi oqadigan bo'shliqni o'z ichiga olgan silindr blokining bir qismidir.*

5 Conclusion

Due to the high importance of foreign languages in the development of the transport sector, it is necessary to train skilled translators within the English-Uzbek framework. Working in cooperation with experts in a specific field and their advice, books, textbooks, and scientific articles makes it possible to provide optimal options in the translation of lexical units in the field of transport (i.e., terms with special words). We need to strengthen cooperation between experts in the field and linguists who can fully analyze linguistic aspects in developing and implementing the system of field-oriented subjects (ESP) of students, who teach translation skills and prepare field translators.

References

1. Averbukh KY (2006) General theory of the term: Textbook. State University of Education, Moscow, Russia
2. Temmerman R (2000) Towards new ways of terminology description: the sociocognitive approach, John Benjamins Publishing Company, Philadelphia, Amsterdam, Netherlands, PA
3. Dabyltaeva RE, Jamaldinova MT (2015) Definition of the term: linguistic and translation aspect. Bull Abai KazNPU Ser: Pedagog Sci 4(48):131–136
4. Tatarinov VA (1992) Terminological views of G.O. vinokur and A.A. reformatsky philological sciences. Sci Essay High Educ 5–6:63–75
5. Dadaboev HA (2016) Modern Uzbek lexicography and terminology. ToshDO'TAU University, Tashkent, Uzbekistan
6. Danilenko VP (1977) Russian terminology: The experience of linguistic description. Nauka, Moscow, USSR
7. Leychik VM (2009) Terminology: Subject, methods, structure, 4th edn. Librocom, Moscow, Russia
8. Newmark P (1988) A textbook of translation. Shanghai Foreign Language Education Press; Prentice Hall International, Shanghai, China
9. Novikov LA (1982) Semantics of the Russian language. Higher School, Moscow, USSR
10. Serbinovskaya NV (2008) *Formation and development of the terminological field "marketing"* (Synopsis of Dissertation of Candidate of Philological Sciences). Southern Federal University, Rostov-on-Don, Russia
11. Proshina ZG (2008) Theory of translation (English and Russian), 3d edn. Far Eastern University Press, Vladivostok, Russia
12. Bobyreva NN (2012) Translation of texts for special purposes: Textbook. Kazan Federal University, Kazan, Russia
13. Kadyrbekova D (2016) Linguistic pecularities of IT technology of English and Uzbek languages. Tashkent, Uzbekistan: Science and Technology
14. Goodsell D (1989) Dictionary of automotive engineering. Society of Automotive Engineers. Butterworths, Warrendale, PA, London, UK
15. Normirzaev A, Mannonov J (2017) Technical operation of automobiles. Namangan, Uzbekistan

16. Akhmanova OS (1966) Dictionary of linguistic terms. Soviet Encyclopedia, Moscow, USSR
17. Prokofyeva EY (2008) Automobile industry formation in Russia (1896–1914). Vestnik Samara State University of Economics 12(50):84–88. https://vestnik.sseu.ru/lk_file.php?pdf=2157. Accessed 19 Feb 2024

Contribution of Tourism in India's GDP in Pre- and Post-pandemic Scenarios

Karan Singh Rana and Roman V. Manshin

Abstract The research investigates the profound impact of the COVID-19 pandemic on the contribution of the tourism sector to India's GDP. Before the onset of the pandemic, tourism was a cornerstone of India's economy, showcasing steady growth and accounting for a substantial portion of national income. The sector generated significant revenue and provided millions of jobs, fostering economic development and cultural exchange. However, the outbreak of COVID-19 brought about unprecedented challenges, resulting in a dramatic decline in tourist arrivals and an associated drop in revenue. Lockdowns, travel restrictions, and health concerns led to widespread cancellations of bookings, severally crippling businesses reliant on tourism. Consequently, these resulted in massive job losses and financial stability for countless individuals and families. This research aims to explore the multifaceted ways in which the pandemic disrupted tourism's economic role in India, emphasizing the urgent need for targeted recovery initiatives. Strategic measures (e.g., promoting domestic tourism, enhancing health and safety protocols, and leveraging digital technology) are crucial to revitalizing the industry. By implementing these initiatives, India can restore the significance of tourism as a vital driver of economic growth and resilience in the post-pandemic landscape.

Keywords Economic growth · Pre-pandemic · Post-pandemic · Tourism industry · Gross Domestic Product (GDP)

JEL codes Z3 · Z320 · O1 · O190 · O2 · O210

K. S. Rana (✉) · R. V. Manshin
RUDN University, Moscow, Russia
e-mail: karanrana388@gmail.com

R. V. Manshin
e-mail: manshin@list.ru

S. M. Sultanova et al. (eds.), *Sustainable Development of Transport*, Lecture Notes in Networks and Systems 1366, https://doi.org/10.1007/978-3-031-88846-5_25

1 Introduction

Tourism economics is a relatively new term in international trade. Currently, tourism is globally recognized as one of the most important sources of foreign exchange earnings for all governments. This industry, as an economic, social, and cultural activity, has become more important to national growth. Tourism is something that, when managed well, has the potential to increase productivity, improve livelihoods, improve public welfare, and use factors such as employment, wealth, and land. Tourism is now the world's largest service sector. Its growth will bring significant social and economic changes [1].

Tourism has played a significant part in the economic development of many countries, including India. Every year, millions of tourists flock to India, drawn by its scenic splendor, historical sites, and rich cultural heritage. Indian tourism significantly contributes to economic growth and prosperity. India is a land of great contrasts, featuring deserts and rainforests, majestic mountains, beautiful valleys, and unspoiled beaches. It caters to nearly all holiday preferences, including adventure travel, beach vacations, ecotourism, and more [2].

The travel and tourism industry can be regarded as India's primary economic growth engine. Travel and Tourism Council (WTTC) estimates that the industry's contribution to India's GDP increased at an impressive CAGR of 6.03% between 2012 and 2018. In terms of travel and tourism (T&T), India ranked eighth globally when it contributed 9.2% of its GDP to the industry in 2018, further to being the 0.33-largest source of foreign exchange profits for the USA; the enterprise hired 8.1% of the staff in 2018. This makes it one of the top process creators in the country [2].

The journey and tourism sector contributed $121.9 billion to the GDP in 2020.

2 Materials and Method

The methodology of the scientific topic is based on traditional methods of economic analysis, such as deduction, induction, comparison, and retrospective analysis. From the point of view of theoretical approaches, the problems of the tourism sector in the formation of GDP have been widely studied by De and Borah [3], Narain and Pandey [4], Nayak and Hanagodimath [5], Shaharban [6], Singh and Alam [7], Sivaprasad [8].

3 Results

3.1 Percentage of GDP is Made from Tour and Tourism

Employment and spending on travel and tourism are measured as a percentage of the overall economy, as reflected in national income accounts and labor market indicators. Tourism-related exports are compared to total exports of all goods and services to assess the sector's contribution. Additionally, expenditures on domestic travel and tourism are evaluated in relation to the country's GDP. The analysis also includes comparing the proportion of government investment in travel and tourism to the total government investment across all sectors.

Tourist and travel consumption is defined in two ways—as spending of residents of the country on tourism within the country and as spending of total expenditure to total domestic expenditure plus total export (Fig. 1).

As shown in Fig. 1, India is recognized as one of the most tourist-friendly nations globally, making the hospitality and tourism sector a key growth driver for the country's service industry. Like many other nations, tourism significantly contributes to India's foreign exchange earnings, underscoring the sector's importance to the Indian economy. From 2016 to 2019, the tourism industry achieved a compound annual growth rate (CAGR) of 7%. However, in 2020, the COVID-19 pandemic led to a decline in tourism earnings.

The travel and hospitality sector, which encompasses lodging and food services, is well known for being a significant source of foreign exchange earnings for many countries, as well as a development and catalyst for socio-economic growth. Tourism showcases India's magnificent history, culture, and variety and generates substantial economic advantages. The travel industry is heeding the call to return to pre-pandemic normality, with federal and state governments dedicated to returning the industry to normal operations [10].

Fig. 1 Overall contribution of travel and tourism to GDP in India from 2013 to 2023 ($ billion). (*Source* Compiled by the authors based on [9])

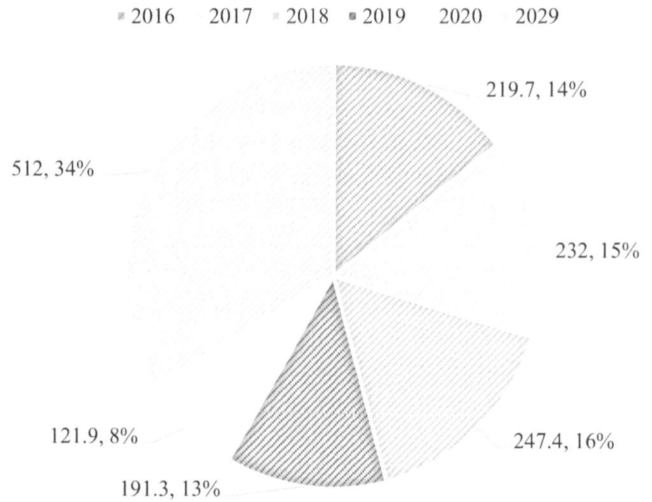

Fig. 2 Total contribution of travel and tourism to GDP ($ billion) (*Source* Compiled by the authors based on [10])

3.2 The Size of the Tourism Industry's Market Considering India

In 2019, India ranked 10th out of 185 nations based on the GDP share of the travel and tourism industry, according to the data provided by the World Travel and Tourism Council (WTTC). Tourism contributed 6% of all economic activity to the economy. Regarding the expenses for the 2019–2020 fiscal year, corporate taxes came to 1,368,100 crore ($194 billion), representing 8% of the total corporate taxes for the year. Additionally, the tourism industry accounted for 78% of India's GDP, totaling 7 billion rupees. Regarding employment, data from 2020 indicates that approximately 39 million Indians were employed in the tourism sector, which constitutes about 8% of the nation's total workforce [1]. Figure 2 shows the total contribution of travel and tourism to the GDP of India.

3.3 COVID-19's Impact on Indian Tourism

The pandemic unprecedentedly impacted the minds of travelers. It created a fear that cannot be removed with or without a vaccine. However, if we are to talk about the sector's revival, certain fundamental adjustments must be made in the way the business operates in many areas of service. The sector has suffered significant losses due to low revenue generation. The travel-tourism industry (TTI) contributed $178 billion to the GDP in 2021. This constitutes one of the serious, harsh challenges facing the Indian tourism industry. As knowledge of the virus spread from November, the percentage of cancellations in this sector increased dramatically, reaching a high of about 80% in March in numerous Indian locales. The value of risk in this category will be multiples of tens of thousands of crores [11–13].

Spreading of the virus must be severely resisted by instilling methods that make the traveler feel safe without being reckless; nonetheless, specific circumstances must be satisfied [1].

3.4 Contribution of Tourism in the Post-pandemic Era

India's tourism and hospitality sector is forecasted to attract over $59 billion by 2028. By the end of 2028, 5 million international visitors are expected to visit the country. According to predictions, India's GDP contribution is estimated to reach $512 billion by 2028. The industry's direct GDP contribution in India is expected to rise by 7–9% each year between 2019 and 2030. According to Statista, the country received more than six million foreign tourist arrivals in 2022, contributing to foreign exchange earnings of over $16.93 billion. Travel and tourism have accounted for Rs. 15.7 trillion in India's economy by 2022 [2, 14].

By 2028, it is anticipated that there will be more than 30.5 million foreign tourists, bringing in almost $59 billion in revenue. It suggests that the expansion will likely be driven by local tourists following the outbreak. It is estimated that the industry will contribute about 47% to the Indian GDP by 2020 and 50% by 2022 due to the increasing growth of international hotel chains in India [15].

According to the Ministry of Tourism, the sector is experiencing a robust annual growth rate of 132.5%. Additionally, FTAs increased by 5%, rising to 795,827 in March of the following year compared to 342,308 in March of the current year. FTAs grew by 222.4% for January-March 2023 compared to the same period in 2022 but decreased by 20.4% by July 10, 2023. In January-March 2019, FTAs increased by 7%. The actual FTA figures were $784,750 for January-March 2022 and $3,179,792 for January–March 2019 [1].

According to the report by news agency ANI, FTAs to India in 2024 are projected to exceed pre-pandemic levels. Industry experts suggest that India is becoming a favored destination for travelers due to the rising conflicts in West Asia and the shift of foreign tourists away from China following the COVID-19 pandemic.

3.5 A Few Initiatives the Indian Government has Taken to Increase Tourism in the Nation

The Ministry of Tourism has declared 2023 as "Visit India Year" to encourage foreign travel, aligning with India's G20 Presidency and the India@75 Azadi ka Amrit Mahotsav celebrations.

The Indian Prime Minister, Narendra Modi, says he had a plan for a mission-mode approach to tourism development during Amrit Kaal, which included a 25-year strategic plan. When we dream about the India 100 in 2047, we want India to be the number one tourist destination for people of the entire globe.

To highlight trade and investment prospects, the Ministry of Tourism, Central Ministries, and State Governments are organizing a Global Tourism Investors Summit. International investors will be able to examine Indian tourism offerings at the event.

Over the years, the Ministry of Tourism has implemented several policies and programs to support and grow the TTI in the nation. Some significant projects are as follows:

- The Union Budget for 2023–2024 allocates $290.4 million to the Ministry of Tourism. Of this amount, $170.85 million is designated for the Swadesh Darshan Scheme, aimed at developing 50 tourism destination packages that will feature physical, digital, and virtual access to "Visit India," along with tourist guides and visitor protection services.
- The Ministry of Tourism has approved 76 projects across 13 separate themes at a revised cost of $668 million. As of December 31, 2021, approximately $95 million of the $534.9 million has been disbursed.
- In the 2023–2024 budget, the government has set aside $30.25 million to find and build PRASAD tourist circuits. Since the scheme's commencement in January 2015, the Ministry has authorized 37 projects in 24 states at a total cost of $146.4 million. A total of $91.6 million has already been spent on these projects. By the end of March 2022, 68 places in 30 states and union territories have been designated for development under the PRASAD Scheme.
- The Ministry of Tourism has worked with the Quality Council of India (QCI) on the SAATHI initiative, which stands for "System for Assessment, Awareness, and Training for Hospitality Industry," to ensure that the hospitality sector can resume operations safely and mitigate risks during the COVID-19 pandemic.
- Under the Loan Guarantee Scheme for Covid-Affected Tourism Service Sector (LGSCATSS), up to $12,110 will be given to tour operators that fall under the Ministry of Tourism and need financial support.
- IBA member banks provide guarantees for skill development loans up to $1801 to qualified borrowers without collateral or third-party guarantees.
- Hunar se Rozgar' Programme, which began in 2009–2010, aims to provide economically disadvantaged youth (originally aged 18–25, later extended to 28) with marketable skills to address the skills gap in the hospitality and tourism sectors and provide financial benefits to the poorest members of the travel industry.

- Prashad and Swadesh Darshan aims to promote interest in sectors such as religious/heritage, wellness/spa, adventure, medical/health, MICE (meetings, incentives, conferences, and exhibitions), wildlife, and others. Additionally, promotional initiatives like the Incredible India 2.0 campaign by TAGS Performance have acted as catalysts for the growth of the tourism industry [16].
- Visa upon arrival: Given the significance of visa policies in augmenting tourism, a trial version of "Long Term Tourist Visas," lasting five years and allowing for multiple entries with a ninety-day visitation period, has been made available to citizens of the eighteen countries that have been chosen.
- Electronic-tourist and medical visa: The government recognized the value of foreign visitors and established the idea of electronic-tourist and medical visa, which has increased the number of foreign visitors to the nation. Tourist security has been enhanced by additional programs, including the multilingual, round-the-clock Atithi Devo Bhava Tourist Helpline.

3.6 Investments and Development in Indian Tourism to Boost the Indian Economy

1. The Ministry of Tourism received an investment of $289.89 million because the industry offers numerous job and business opportunities for young people.
2. As per Section 35AD of the Income Tax Act, newly listed hotels in India that are rated two stars or above are eligible for a 100% deduction for all capital expenditures.
3. In 2019, the Indian government lowered the GST rates to 12% for all hotel rooms, which fell between $14.32 and $107.31. This move was made to increase tourism in India.
4. The Ministry of Tourism was allocated $290.64 million in the Union Budget 2023–2024, recognizing the sector's ability to provide numerous employment and startup opportunities. Additionally, the Ministry aims to promote tourism through active state participation, the integration of government initiatives, and public-private partnerships.
5. As per our plan, we are going to develop an app that would have all the features related to travel. The states will be encouraged to open Unity Malls in their respective capital cities, as these are the most popular tourist destinations. The international market will feature One District One Product, GI products, handmade products, and products from neighboring states.
6. To ensure that incentive coverage is distributed globally in line with the distribution of source markets, the Government of India announced in early 2021 that out of 500,000 free ordinary visas, only 40,000 would be e-tourist visas for tourists.
7. The Indian government plans to build 220 additional airports in 2025.
8. Businesses in the tourism and hospitality sector currently qualify for the ECLGS scheme because it encompasses a wide definition of small and medium-sized

enterprises. Exhibition/convention centers of medium and large sizes have been given infrastructure status.

9. A special liquidity facility of $1.8 billion has been made available for the industry.
10. FDI inflows into the tourism and hospitality sector stood at $16.48 billion between April and June 2022.
11. As of September 2022, 11,220 lodging units had self-certified for SAATHI requirements out of the 48,775 lodging units (including classified and unclassified) registered on the National Integrated Database of Hospitality Industry (NIDHI) site.
12. In October 2022, the Indian Hotels Company (IHCL) announced the opening of Loya, a new restaurant brand with an Indian theme. Loya, which made its debut at the Taj Palace in New Delhi, embodies the essence of North Indian cuisine.
13. In May 2022, the hospitality startup OYO paid $5.5 million to buy the European vacation rental startup Direct Booker.
14. Accor of France, continuing its expansion in the Asia-Pacific region and particularly in India, is expected to open nine additional hotel properties in the country, bringing the total number of operated Accor hotels in India to 54. All of these real estate projects are in the economy and mid-market segments.
15. The global medical tourism industry is projected to grow from 2020 to 2027 at a compound annual growth rate (CAGR) of 21.1%.

4 Conclusion

India offers a wealth of tourism opportunities that can serve as a key to unlocking long-term economic growth, increasing employment, and developing infrastructure. This potential is particularly evident in sectors such as specialty travel and cultural heritage tourism, which contribute to raising living standards and advancing the economy through goods and services.

In the 1950s, the Indian government and other relevant authorities recognized that developing tourism could be a viable alternative for the nation's economic growth. However, the government did not focus on tourism development until 1966, when it introduced regulations supporting the India Tourism Development Corporation. These regulations enabled the establishment of several five-year programs aimed at promoting tourism.

The COVID-19 pandemic posed a significant threat to the global tourism industry, leading to widespread shutdowns. However, as the tourism sector recovers from the pandemic, it is expected to return to pre-pandemic levels due to increased foreign tourist visits and spending. From now through 2024 and in the subsequent years, India is poised to expand its tourism sector and reap its associated benefits.

References

1. Ministry of Tourism Government of India (2023) India: tourism statistics at a glance—2023. New Delhi, India. https://tourism.gov.in/sites/default/files/2023-07/India%20Tourism%20Statistics%20at%20a%20glance%202023%20-%20English%20version_0.pdf. Accessed 2 Apr 2024
2. Vamshidhar M, Sreyesh S (2024) A comprehensive study on the macro-economic impact of tourism in India. Indian Soc Tech Educ 47(1):76–82
3. De UK, Borah B (2024) Impact of COVID-19 pandemic on tourism in India. J Tour Sport Manag 4(3):1140–1150
4. Narain A, Pandey B (2023) A critical analysis of tourism and hotel industry: with special reference to India. Int J Creat Res Thoughts 11(8):g641–g654. https://www.researchgate.net/profile/Aditi-Narain/publication/380743629_A_CRITICAL_ANALYSIS_OF_TOURISM_AND_HOTEL_INDUSTRY_WITH_SPECIAL_REFERENCE_TO_INDIA/links/664cacdb479366623a050b70/A-CRITICAL-ANALYSIS-OF-TOURISM-AND-HOTEL-INDUSTRY-WITH-SPECIAL-REFERENCE-TO-INDIA.pdf. Accessed 2 Apr 2024
5. Nayak J, Hanagodimath SV (2024) An impact analysis of tourism and economic growth in India. Int Educ Res J 10(8):44–49. https://doi.org/10.21276/IERJ24502596312007
6. Shaharban V (2024) Tourism industry in India after COVID-19 pandemic: exploration of the latest trends. Int J Res Anal Rev 11(3):426–431. https://doi.org/10.2139/ssrn.4960708
7. Singh D, Alam Q (2024) Is tourism expansion the key to economic growth in India? an aggregate-level time series analysis. Ann Tour Res Empir Insights 5(2):100126. https://doi.org/10.1016/j.annale.2024.100126
8. Sivaprasad KI (2024) Role of the tourism sector in carving out a better India. EPRA Int J Environ Econ Commer Educ Manag 11(1):5–9. https://doi.org/10.36713/epra15458
9. Statista (2023) Contribution of travel and tourism to GDP in India from 2013 to 2023. https://www.statista.com/statistics/313724/total-contribution-of-travel-and-tourism-to-gdp-in-india-by-segment/. Accessed 2 Apr 2024
10. India Brand Equity Foundation (IBEF) (2022) Tourism and hospitality industry in India: market size, government initiatives, investments. https://www.ibef.org/industry/tourism-hospitality-india. Accessed 12 Jul 2024
11. Hindustan times (2023) Foreign tourist arrivals to India in 2024 to surpass pre-pandemic level: Experts. https://www.hindustantimes.com/world-news/foreign-tourist-arrivals-to-india-in-2024-to-surpass-pre-pandemic-level-experts-101714385271672.html. Accessed 12 Jul 2024
12. Pandey K, Mahadevan K, Joshi S (2021) Indian tourism industry and COVID-19: a sustainable recovery framework in a post-pandemic era. Vis: J Bus Perspect 28(5):592–606. https://doi.org/10.1177/09722629211043298
13. The Hindu Bureau (2023) Tourism to contribute $512 bn to India's GDP by 2028, says TAAI chief. https://www.thehindu.com/business/tourism-to-contribute-512-bn-to-indias-gdp-by-2028-says-taai-chief/article66468359.ece. Accessed 2 Apr 2024
14. India Brand Equity Foundation (IBEF) (2023) The travel and tourism sector is expected to contribute US$ 201.37 billion (Rs. 16.5 trillion) to India's economy by 2023: world Travel and tourism council. https://tinyurl.com/yc7nmxsc. Accessed 12 Jul 2024
15. Alluri S, Venkateswarlu P (2023) A review on pre- and post-COVID-19 scenario of Indian tourism. J Humties Soc Sci 6(8):282–284. https://www.opastpublishers.com/open-access-articles/a-review-on-pre-and-post-covid19-scenario-of-indian-tourism.pdf. Accessed 2 Apr 2024
16. Ministry of Tourism, Government of India (2023) India tourism statistics 2023. https://tourism.gov.in/sites/default/files/2024-02/India%20Tourism%20Statistics%202023-English.pdf. Accessed 2 Apr 2024

ESG Agenda of the Future and the Prospects for Sustainable Transport Development in Accordance with It (Conclusion)

This first volume of the book unveiled the scientific understanding of the significance that the contemporary ESG agenda holds for transport and logistics systems and how to implement this agenda within these systems to achieve sustainable development. The most important conclusions drawn from the research conducted in this volume are as follows.

First, managing the sustainable development of territorial transport and logistics systems should be conducted in accordance with the ESG agenda. This requires a green transition in transport logistics, which involves the environmentalization of transport and logistics services through waste minimization and a shift to renewable energy sources, as well as the greening of logistics facilities in urban areas.

Second, digital technologies and infrastructural support are crucial for the sustainable development of transport and logistics systems. The extensive opportunities for digital modernization in transport logistics are paving the way for creating high-tech and eco-friendly clusters in this sector. This volume proposes practical solutions for the sustainable development of transport logistics in smart cities, such as smart parking systems and Logistics 4.0.

Third, social communications within the transport logistics system are critically essential for its sustainable development in line with the ESG agenda. The network of social connections in transport logistics includes the training of personnel for transport organizations, the management of human resources within these organizations, and their marketing interactions with stakeholders, including representatives of local communities in the cities where transport and logistics services are provided.

In addition to scientific discoveries and developments, the second volume of the book has raised new research questions that merit attention and scholarly exploration. Similar to the Sustainable Development Goals (SDGs) that replaced the Millennium Development Goals, the ESG agenda is subject to review and modernization as the global economy evolves. The current ESG agenda is focused on a Decade of Action. After 2030, it may undergo significant changes.

S. M. Sultanova et al. (eds.), *Sustainable Development of Transport*, Lecture Notes in Networks and Systems 1366, https://doi.org/10.1007/978-3-031-88846-5

From an environmental (E) perspective, the SDGs call on the global community to reduce the economic carbon footprint and decrease production and consumption waste by 2030 while aiming for complete carbon neutrality of the economy by 2050. To support this goal, transport logistics must transition to climate-smart technologies that accelerate the green shift and fully adopt clean energy sources.

From a social (S) perspective, globalization is taking on new forms; its contours may change significantly after 2030. Based on current experience, a further increase in demand for digital and green personnel in transport organizations can be expected. Additionally, new logistical routes may be required in the future, necessitating marketing support—opportunities and directions for which should begin to be developed now.

From a corporate governance (G) standpoint, the emergence of innovative governance tools is anticipated. The role of science in this context is to identify the advantages and disadvantages, as well as the limitations, of using these emerging tools in the corporate governance of transport organizations to ensure their sustainable development.

Thus, the conceptualization of the future ESG agenda, the determination of its prospects, and the development of recommendations for sustainable transport development in accordance with it should be the focus of further scientific research, continuing the work presented in this second volume of the book.

The manufacturer's authorised representative in the EU is Springer Nature Customer Service Centre GmbH, Europaplatz 3, 69115 Heidelberg, Germany. If you have any concerns regarding our products, please contact ProductSafety@springernature.com

Printed and bound by CPI Group (UK) Ltd, Croydon, CR0 4YY
01/10/2026
02225966-0007